THE EARTH'S MAGNETIC FIELD

The Earth's Magnetic Field

William Lowrie

OXFORD

UNIVERSITY PRESS

OXFORD
UNIVERSITY PRESS

Great Clarendon Street, Oxford, OX2 6DP,
United Kingdom

Oxford University Press is a department of the University of Oxford.
It furthers the University's objective of excellence in research, scholarship,
and education by publishing worldwide. Oxford is a registered trade mark of
Oxford University Press in the UK and in certain other countries

Published in the United States of America by Oxford University Press
198 Madison Avenue, New York, NY 10016, United States of America

British Library Cataloguing in Publication Data
Data available

Library of Congress Control Number: 2022952150

ISBN 978–0–19–286267–9
ISBN 978–0–19–286268–6 (pbk.)

DOI: 10.1093/oso/9780192862679.001.0001

Printed and bound by
CPI Group (UK) Ltd, Croydon, CR0 4YY

Foreword

The branch of geophysics that specializes in describing and investigating the Earth's magnetic field is called geomagnetism. It comprises many subdisciplines. Like many other scientists, I spent my research career in quite a narrow field of geomagnetism, and, although I was aware of major advances in the other fields, my research did not stray far from familiar territory. Writing academic books that embrace a range of topics, especially since I retired, has therefore been very educative. Preparing background material for the present book has made me familiar with some remarkable achievements in many research fields within the framework of geomagnetism. They range from measuring the Earth's magnetic field with remarkable precision from satellites that orbit the planet hundreds of kilometers above its surface to simulating with supercomputers the complex processes in the Earth's core that generate the field. Understanding the Earth's magnetic field also requires becoming familiar with the magnetic field of the Sun, the flow of energetic particles radiated by it, and their many interactions with the geomagnetic field. I hope the reader will be inspired to investigate some of these topics in more depth than can be presented in this introductory text.

The geomagnetic field is a complex and important property of the Earth. The physical presence of the field is vital for our planet's habitability, including existence of life on it. Physicists use mathematical methods to explain phenomena. However, modeling mathematically the processes by which the geomagnetic field is generated is one of the most difficult tasks in theoretical geophysics. The equations must simultaneously describe the thermal conditions in the Earth's core, the motions of the electrically conducting molten iron that forms the liquid outer core, the effects of the Earth's rotation, and electromagnetic interactions that are involved in generating a magnetic field that is able to sustain itself. The equations cannot be solved analytically. Our understanding of how the field originates is obtained from sophisticated computer models, which are able to simulate the dynamo mechanism that generates the field and to compute how it behaves with time, including its ability to reverse polarity. Although the general features of the process are well understood, many details have yet to be worked out.

Only a comparatively small group of theoretical geophysicists have the mathematical expertise to investigate the intricacies of what is happening in the Earth's core, but many geophysicists with other specializations are active in studying different aspects of the geomagnetic field and how it can be used for the benefit of society. Space physicists study the Sun and its emissions, applied geophysicists make use of magnetic anomalies to interpret structures in the Earth's crust, and geologists use the record of ancient magnetic fields preserved in rocks to untangle the planet's geology. From outer space to the central core, geomagnetism is a lively field of current research.

In this book I have tried to present the main features of the Earth's magnetic field in a way that is understandable to the nonexpert reader. I have avoided the derivation of the underlying equations, and only a small number are displayed for the purpose of clarifying explanations. In my attempts to describe the various aspects of geomagnetism, I received advice and suggestions from several friends. In particular, Chris Finlay, Alan Green, Ann Hirt, Walter Hirt, Dennis Kent, and Alexei Kuvshinov kindly helped me by reviewing draft versions of individual chapters. I thank them for their criticisms and corrections and for taking time from their main activities to help me improve the manuscript. I also thank an anonymous reader for useful comments and corrections. As always, I am grateful to my wife Marcia for her constant support and encouragement, and for reading and criticizing the text. The book is dedicated to Marcia.

Contents

List of Illustrations

1

What is Magnetism?

Introduction

Since the development of simple compasses, the Earth's magnetic field has been an important aid for navigation, enabling travelers to find their way in uncharted regions. It became an essential tool in the exploration of the planet and facilitated the great voyages of discovery in the late 15th century that established links between the continents. It is one of the central properties of the Earth and has been studied with purpose-built equipment on land, at sea, and in the past half-century from satellites. The magnetic field extends far into space around the planet, where it interacts with the energetic particles from the Sun and interstellar space that impact on the Earth. It produces sideways forces on the rapidly moving, electrically charged particles and deflects them around the planet. In this way, the magnetic field acts as a protective shield against harmful radiation and enables life to exist on the planet.

Geomagnetism, as the study of the Earth's magnetic field is formally called, began with simple observations of natural phenomena. For centuries these phenomena were assumed to have celestial origins. Gradually, knowledge of the field developed, and it became recognized as a property of the Earth itself. Laboratory investigations in the early 19th century established the relationship between electricity and magnetism. Subsequently, our knowledge and understanding of the field have grown almost exponentially. It is now known that the geomagnetic field results from electrical currents in the Earth's interior, in its atmosphere, and in space around it. However, explaining the mechanism by which the field is generated depends on thermal processes and material properties in the Earth's core that are still incompletely resolved. Moreover, the geological history of the field, as well as its relationship to dynamic changes in the Earth's interior and at its surface, are poorly known for most of the planet's 4,500 Myr existence.

1.1   The Discovery of Magnetism

The existence of magnetic forces has been known since long before the beginning of the Christian era; they were discovered by chance and were not investigated systematically.

The Earth's Magnetic Field. William Lowrie, Oxford University Press. © William Lowrie (2023).
DOI: 10.1093/oso/9780192862679.003.0001

The reports of early discoveries of magnetic properties were written down centuries after the event and may be more legend than fact. The Greek and Chinese civilizations of antiquity were familiar with naturally occurring stones that displayed strange behavior. The stones were attracted to each other and to iron implements. According to some sources, the word "magnet" is associated with a region of Ancient Greece called Magnesia, located in what is now western Anatolia, where the properties of these unusual stones were first observed. The Roman historian known as Pliny the Elder relates a different legend of Ancient Greece, in which a shepherd called Magnes noticed that the iron studs of his boots seemed to stick to certain stones. Scientific analysis shows that the stones, which from the Middle Ages have been called lodestones, contain intergrowths of the minerals magnetite and maghemite, which are strongly magnetic oxides of iron. Only a few naturally occurring minerals are sufficiently magnetic to be able to attract other magnetic materials, so the magnetism of lodestone is a rare property. In order to become magnetized, a material must be exposed to a magnetic field. It is possible that a lodestone acquires strong magnetization as a result of lightning strikes.

Because there was no basis for understanding their behavior, lodestones were regarded with wonder, and their properties were used in geomancy—the art of interpreting divine intention by studying the pattern formed by rocks or pebbles when cast upon the ground. A practical use for magnetism was developed by Chinese natural philosophers, who discovered its orienting capability. They observed that a lodestone suspended from a thin thread would adopt an alignment with the north–south direction, so that it could be used for navigation when the Sun or stars were not visible.

The ability of a lodestone to pass on its magnetic properties to iron objects was also known in antiquity. A dialogue between Socrates and his pupil Ion, written in the year 380 BC by Plato, makes reference to "a stone which Euripides calls a magnet" that "not only attracts iron rings, but also imparts to them a similar power of attracting other rings" (from Plato's Dialogues: Ion, trans. Benjamin Jowett, 1871). Thus, after stroking an iron needle with a lodestone, the needle itself became a permanent magnet. This observation was an important scientific advance. An iron needle magnetized in this way could easily be suspended by a thread or floated on water on a piece of wood or cork. An iron needle was more strongly magnetic than a lodestone and therefore acted as a more sensitive indicator of the north–south direction.

The magnetized needle gradually became used as an elementary compass. Simple magnetic compasses were in use by Chinese navigators late in the first millennium of the current era and were subsequently introduced to Europe by way of the Arab world. However, the nature and origin of the forces that oriented the compass needle were not at all understood. Until the 12th century, the magnetic alignment was believed to be controlled by the heavens, which worked in such a way that the compass needle was attracted by some unknown means toward the pole star.

Magnetic compasses were adapted for efficient use on ships, where they were mounted in gimbals to accommodate the pitch and roll of the vessel. The maritime compass made possible the great expeditions of the European adventurers who discovered and explored the Pacific and Atlantic oceans and the New World. During their travels in the 15th and 16th centuries, mariners observed carefully the direction of the compass needle and noted small but systematic differences. As a result, the historic logs of ships' positions provide archives of the magnetic field directions at that time. With the aid of modern computers, scientists have been able to reconstruct the geometry of the Earth's magnetic field from the present time back to the 16th century (Jackson, Jonkers, and Walker, 2000).

In scientific terms, the magnetic field is a vector quantity: that is, it has both intensity (strength) and direction. The intensity of the field decreases with increasing distance from the source, and it also varies with azimuth around the source. The strength and direction of the field can be described at any location on the Earth's surface by expressing it as three components at right angles to each other (Fig. 1.1). The north (X), east (Y), and vertically downward (Z) directions form a cuboid, with the total field (F) as its diagonal. The north and east components define the horizontal component (H), which is the direction to the north magnetic pole. The angle between magnetic north and geographic north is the declination (D). The horizontal and vertical components define a vertical plane, the magnetic meridian, in which the angle between the total field (F) and the horizontal component is called the inclination (I). Each component varies from one location to another and also changes on a wide spectrum of time scales, ranging from fractions of a second to millions of years.

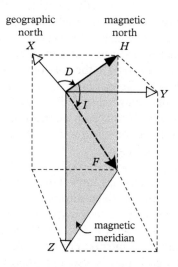

Fig. 1.1 *The geomagnetic field elements consist of three orthogonal components (X, Y, and Z), which define the total field (F), horizontal component (H), and the angles of inclination (I) and declination (D).*

1.2   The Earth as a Magnet

Several early pioneering studies contributed to understanding the magnetic field and the idea that it is a property of the Earth. In 1269, a medieval French scholar, known as Petrus Peregrinus de Maricourt, wrote a letter (called in short form *Epistola de Magnete*) to a friend in which he described experiments he had carried out on the magnetic properties of iron needles and lodestones. He observed that the magnetism of a sphere of lodestone, which he called a *terrella* (or little Earth), varied over its surface but was concentrated at diametrically opposite points. He used the term *pole* (from the Latin word *polus*, meaning the end of an axis) to refer to these centers of magnetic force, and he noted that there were different types of poles at the opposing ends of the magnetic axis. He established that magnetic poles of the same type repel each other, whereas unlike poles attract each other. However, instead of recognizing the ability to align in the north–south direction as an intrinsic property of the magnet, he attributed it to a celestial cause, such as the influence of Polaris, the pole star. This notion was to persist for almost three centuries. There was little further advance in the understanding of magnetism or the magnetic field during the Middle Ages that followed.

The north direction indicated by a magnetic compass and the true geographical north direction are not quite identical. The discrepancy is called the *declination* of the field (Fig. 1.1). The earliest record, in 1088, of the angle of declination is attributed to Shen Kuo, a Chinese intellectual who was knowledgeable in many subjects. The declination arises because the magnetic poles do not exactly coincide with the geographic poles. The vertical plane that passes through the horizontal component of the field intersects the surface of the Earth on a great circle called the *magnetic meridian*. In 1576, Robert Norman, an English compass-maker, showed that a magnetized needle that is free to swing in the meridian plane aligns at an angle to the horizontal. This observation appears to have already been made in 1544 by Georg Hartmann, a German instrument maker, but his result did not become public knowledge until it was discovered centuries later. The angle between the field and the horizontal is called the *inclination* (or *dip*) of the field (Fig. 1.1). It is defined by convention to be positive when the field points downward, as it does in most of the northern hemisphere.

In 1600, William Gilbert, a natural scientist who became the physician to England's first Queen Elizabeth, published a historically important scientific work on magnetism, titled in short form *De Magnete*. In this work he summarized all that was known at that time about the Earth's magnetic field, with due recognition of the earlier work of Petrus Peregrinus on the magnetism of a sphere of lodestone. He also added his own observations based on careful scientific experiments. For example, he showed that when a magnet is divided into two parts, each part is also a magnet, with a pole of attraction at one end and a pole of repulsion at the other end. Most importantly for understanding the magnetism of the Earth, he showed that small magnetized needles adhered to the surface of a lodestone sphere at angles that resembled the known variation of the inclination of the Earth's magnetic field along a circle of longitude. The direction of the field is parallel to the surface of the Earth (i.e., horizontal) at the magnetic equator and perpendicular to it (i.e., vertical) at the poles. Gilbert inferred from his observations that the Earth itself is a giant magnet. This finding was an important step philosophically because—in contrast to Peregrinus—he recognized that the magnetic field is a property of the Earth and not the expression of a celestial force.

It sometimes happens that different scientists may be studying the same phenomenon at the same time and in a similar way. Independently of Gilbert, a French nobleman, Guillaume le Nautonier (William the Navigator), was also interested in the magnetic behavior of the lodestone. In 1603, he published an account in which he concluded that the force that produced the alignment of a compass needle originated within the Earth itself. Both Gilbert and le Nautonier took the bold step of attributing the Earth's magnetic field to an internal property of the planet instead of to a celestial cause. This point of view was not shared by the contemporary Christian church, which regarded many theories that differed from its teachings as heretical. However, these pioneering investigations, based on experimentation rather than speculation, form the beginnings of geomagnetism as a science.

Gilbert imagined the magnetic field at any point on Earth to be constant and invariable with time. This concept was disproved in 1635 by Henry Gellibrand, an English mathematician, who noticed that the angle of declination in London had changed by 7 degrees in 54 years. This slow change of the field with time is called its *secular variation*. It is also observed in other components of the field and is a topic of great interest to modern experts in geomagnetism as its analysis reveals important information about how the field is generated.

Gradually, systematic measurements of the geomagnetic field were made over much of the Earth's surface, so that its spatial variation with position became increasingly well known. For example, in 1701 Edmond Halley—an English astronomer, after whom the comet is named—produced a chart of declination throughout the Atlantic Ocean. To do so, he introduced the technique of interpolating contour lines between the actual measurement positions. The use of contour lines became an indispensable technique in map-making and other scientific analyses. However, it would take more than another 200 years before the origin of magnetism and its relationship to macroscopic and atomic electrical currents were established. This fundamental knowledge derived from laboratory investigations early in the 19th century.

1.3 The Origin of Magnetic Fields

Before proceeding any further, we must define what is meant by the term *magnetic field*. The expression can be understood in two ways. In physics it may refer to the *strength* of a force, in which case it is expressed as the force per unit of the quantity that produces the force. For example, the gravitational field of an object is the force of attraction it exerts on a unit of mass (e.g., kilogram). Similarly, the strength of an electric field is the force it exerts on a unit of electrical charge. Electrical charges can be negative or positive, and the force between two charges is attractive when the charges are of opposite kinds and repulsive when they are of the same kind. The strength of a magnetic field represents a "force per unit of magnetism," but this notion is both vague and inexact because magnetism originates from electrical currents, that is, moving electrical charges.

The other usage of the term *magnetic field* is to describe the *geometry* of the region in which the force acts. Michael Faraday, a pioneering English scientist, introduced the term in 1845 to describe the effect of a magnet by imagining the lines of force surrounding it. The concept can be applied to describe any force field. For example, the direction of an electrical field is defined by its force of repulsion on a positive charge. Thus, electrical field lines act radially inward (attraction) around a negative charge and radially outward (repulsion) around a positive charge. The strength of the field decreases along a field line with increasing distance from the charge, following an inverse-square law, but its direction remains radial.

The magnetic field of a simple magnet has two centers of force. When a fine powder of iron filings is sprinkled on a sheet of paper that lies on top of a straight bar magnet, the magnet's lines of force become visible (Fig. 1.2). Each grain of the powdered iron behaves like a tiny magnet and aligns with the force field of the magnet. The tiny particles tend to clump together, forming ragged but distinct lines that converge on two centers of force, one at each end of the magnet. The lines of force enter the magnet at one end and leave at the other. They spread out in the space between them, outlining clearly the geometry of the force field of the bar magnet.

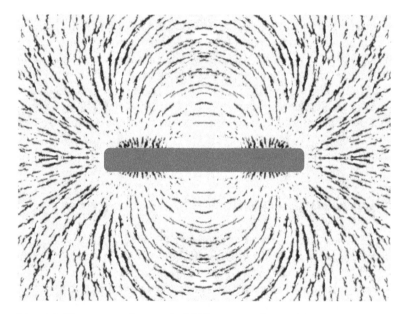

Fig. 1.2 *The pattern of magnetic field lines around a straight bar magnet (the outlined dark block) is revealed by a powder of iron filings sprinkled on a sheet of paper covering the magnet.*

The centers of force are called the *poles* of the magnet, as first named by Peregrinus. They are of equal strength but opposite sign. Each pole exerts a force of repulsion on a similar pole, while the other pole is attracted to it. It has been known since the observations of Peregrinus in the 13th century, reinforced by Gilbert in the 17th century, that when a magnet is divided into smaller parts, each part is observed also to have two poles of opposite sign; that is, magnetic poles always occur in pairs. The ability of a freely suspended magnet to rotate into alignment with an external magnetic field depends on the strength and separation of the magnet's poles and is called its *magnetic moment*. The concept is used to describe the strength of the Earth's field as well as magnetic behavior at an atomic level.

The attraction of the Earth's magnetic field on one end of a compass needle and the repulsion of the other end cause the compass needle to align in the north–south magnetic meridian. By convention, the north-seeking magnetic pole of the compass was associated with the north *geographic* pole, and the place to which it pointed was called the *north magnetic pole*. The locations are in fact quite far apart: the north magnetic pole is currently located in the north of Canada, several hundred kilometers from the geographic pole. The historic labeling of the magnetic poles is unfortunate. In physics, magnetic field lines are defined so that they leave magnetic north poles and return to magnetic south poles. Moreover, the Earth's magnetic field is defined to act northward and downward in the northern hemisphere (Fig. 1.1). Paradoxically, this means that the north pole of the Earth's magnetic field must *physically* be a magnetic south pole. The contradiction arises from associating the magnetic field lines with fictitious magnetic poles.

The classical physics of magnetism is based on the assumption that individual magnetic poles (referred to as monopoles) do not exist; this assumption is borne out by experience. Modern physics has never established the existence of magnetic monopoles, although experimental efforts have been made to find them. Although they are a fictitious concept, magnetic poles can be useful in geophysics. They are, for example, a useful aid to understanding the magnetic signatures of geological structures and orebodies in the shallow subsurface.

A long, thin, magnetized needle—whose poles are far apart and cannot influence each other—can be used to understand the properties of magnetic forces. Each end of the needle approximates a single pole, and the lines of magnetic force act radially to it. The magnetic force decreases with distance from a single pole as the *inverse square* of the distance. However, this is an artificial situation. In practice, a magnet is usually short enough for both poles to affect the shape of its magnetic field, which is described by curved lines that lead from one pole to the other. The decrease of field strength with distance is then more complicated and faster than described by an inverse-square law. The field lines mark the magnet's direction of force; at any point, it acts parallel to the direction of the field line at that point. The geometry of the field lines therefore describes the shape of the magnetic field. The term *magnetic flux* describes the amount of magnetic field that "flows" through a unit of area in a direction normal to its surface. Where the field lines are close together (e.g., near the magnetic poles), the magnetic flux is high; where the field lines are far apart, it is low.

A *dipole* field arises when the poles are so close together that their separation is negligible compared to the distance of the observer from them (Fig. 1.3). At any point in the field of a dipole, the effects of both poles are felt. As a result of this superposition, the strength of the magnetic field decreases as the inverse *cube* of the distance from the midpoint of the dipole. At a constant distance from the dipole (i.e., on a spherical surface around the dipole), the direction of the field varies with the angular distance from the dipole axis. It points vertically inward or outward over the poles of the magnet, and—by analogy to the Earth's field—it is "horizontal" over the "magnetic equator."

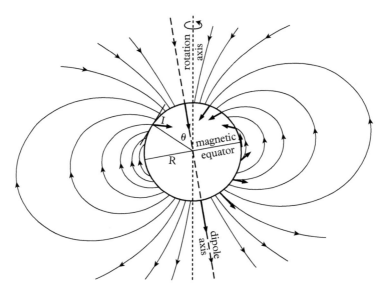

Fig. 1.3 *The field lines of a magnetic dipole at the center of the Earth. The axis of the dipole is tilted at about 10 degrees to the rotation axis. Its magnetic field enters the Earth in the northern hemisphere and leaves in the southern hemisphere. The angle θ is the magnetic colatitude of a position on the surface. The angle I is the inclination of the field at that place. It is horizontal at the magnetic equator and vertical at the magnetic poles.*

The dipole field is the most important and fundamental type of magnetic field. It is, for example, the type of magnetic field around a uniformly magnetized sphere of magnetic material (Fig. 1.4a), such as lodestone, which is why Gilbert's experiments were successful in describing some properties of the geomagnetic field. It is also the shape of the magnetic field produced by an electrical current in a small loop of conducting wire (Fig. 1.4b). At a more elemental level, it is the type of magnetic field produced by an electron orbiting the nucleus of an atom.

1.4   Electrical Currents and Magnetic Fields

An electrical current is a flow of electrically charged particles. In a conductor, it consists of negatively charged electrons; in a liquid, it consists of positively or negatively charged atoms (called ions). A stream of charged particles from the Sun is made up of both electrons and ions. The measure of how easily a material allows an electrical current to pass through it is called its *conductivity*; it is expressed in units of Siemens/meter (S/m). The inverse, that is, how strongly the material opposes a current, is called its resistivity and is measured in ohm meters. The geomagnetic field is produced by various electrical currents, flowing primarily in the liquid core, but also in space around the Earth, in the atmosphere, in the oceans, and in the solid interior of the planet.

The physical relationship between electricity and magnetism was discovered as a result of experiments carried out in the early 19th century. In 1820 a Danish scientist, Hans Christian Ørsted, showed experimentally that a magnetized iron needle was deflected in the vicinity of an electrical current. In experiments that augmented this observation, the French scientist André-Marie Ampère demonstrated that a force of repulsion or attraction existed between two parallel current-carrying conductors, depending on whether the currents were in the same or opposite directions. The law governing the direction and strength of the magnetic field around each conductor was developed by two French physicists, Jean-Baptiste Biot and Félix Savart.

An electrical current produces a magnetic field in a plane perpendicular to the current. For example, a steady current through a straight wire creates a magnetic field around the wire that has the shape of closed circles in a plane perpendicular to the wire. Conversely, a current through a small loop of wire creates a dipole field around it (Fig. 1.4b). The direction of the magnetic field along the axis of the loop is defined by a right-hand rule: with the fingers of the right hand pointing along the direction of the current, the thumb indicates the direction of the magnetic field it produces.

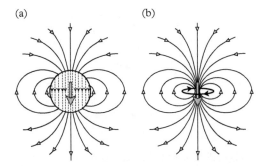

Fig. 1.4 *A dipole magnetic field is produced (a) by a sphere of magnetic material that is magnetized uniformly, as well as (b) by an electrical current that flows in a small conducting loop.*

An important corollary of these observations is that a stationary electrical charge does not produce a magnetic field, which only results when electrical charges are in motion. In a complementary way, a steady magnetic field does not produce an electrical current, but a *changing* magnetic field does. This was shown by the English scientist Michael Faraday, who—although he had little formal education or mathematical training—was a gifted experimenter. In 1831, by moving a magnet into and out of a coil of wire, he demonstrated that a current was induced in the wire only when the magnet was in motion. The process is called electromagnetic induction (or simply magnetic induction). In 1834, a Russian scientist, Emil Lenz, established that the direction of the induced current is such that its own magnetic field opposes the initial change in the magnetic field. The fundamental experimental results obtained by Faraday and Lenz established the fact that electrical currents and magnetic fields are not independent.

An electrically charged particle that moves through a magnetic field experiences a force that acts perpendicular to both the field and the direction of motion. This relationship was established in 1895 by a Dutch physicist, Hendrik Lorentz, after whom the force is named. If the velocity **v** of the electrical charge q makes an angle θ with the magnetic field **B** (Fig. 1.5), the charge experiences a deflecting force **F** given by the Lorentz equation

$$\mathbf{F} = q\,(\mathbf{v} \times \mathbf{B}) \tag{1.1}$$

In the Lorentz equation, the electric charge q is a scalar quantity; it has a magnitude but no direction. In contrast, each of the quantities **F**, **v**, and **B** is a vector with both a magnitude and a direction. The rules by which vectors are combined in mathematical equations need to accommodate their directional properties as well as their nondirectional values. The Lorentz equation is a straightforward example. The multiplication

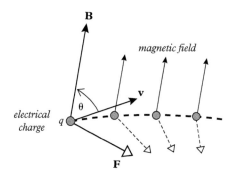

Fig. 1.5 *When an electrical charge* q *moves at velocity* v *through a magnetic field* B, *it experiences a Lorentz force* F *at right angles to both the field and its direction of motion, deflecting it continually so that its path becomes a curved line.*

symbol in Eq. (1.1) stands for the vector-product (also called the cross-product) of the two vectors **v** and **B**. It defines the direction of the deflecting force **F** to be perpendicular to both the velocity and magnetic field at each point of the path of the moving charge. This causes the path to be curved (Fig. 1.5).

The magnitude of the force F is a scalar product:

$$F = qvB \sin \theta, \tag{1.2}$$

where θ is the angle between the vectors **v** and **B** and sin is the trigonometric sine function of the angle. The force on a unit of electrical charge is the electric field and is denoted **E**. Dividing both sides of the Lorentz equation by the charge q leads to the following alternative form of the equation,

$$\mathbf{E} = \mathbf{v} \times \mathbf{B} \tag{1.3}$$

The Lorentz force has important consequences for the interaction of electrical currents and magnetic fields and plays an important role in the definition of the fundamental units of electricity and magnetism. It is a basic principle in the design of strong-field physics equipment such as cyclotrons and other particle accelerators. It is also an important factor in generating the geomagnetic field in the Earth's fluid core.

The Lorentz force also helps us to understand how the geomagnetic field protects the Earth. When an electrically charged particle from the Sun, or distant regions of space, impacts on the Earth, it encounters the surrounding magnetic field. The Lorentz force acts on charged particles and deflects their paths so that most pass around the planet.

1.5   Magnetism at the Atomic Level: The Bohr Model of the Atom

In order to measure the magnetic fields of distant stars or to study the magnetic field at the surface of the Sun, astrophysicists use a method based on the quantum mechanical properties of atoms. The method makes use of the fact that the frequency of the energy radiated by an atom is altered in the presence of a magnetic field. A practical starting point for explaining how this change happens is the Bohr model of the atom, which uses the concepts of classical physics. The model is unable to explain all aspects of atomic behavior and becomes inadequate, for example, when the angular momentum of the electron spin must be taken into account. A profound analysis requires knowledge of quantum mechanics, which is beyond the scope of this book. Nevertheless, it is possible to appreciate some aspects of magnetism at the atomic level without getting deeply involved in this advanced topic.

All magnetic fields arise from electrical currents; even in a permanently magnetized solid, such as a needle or an iron bar, the magnetism results from moving electrical charges. In 1913 a Danish physicist, Niels Bohr, presented a simple model of the structure of the atom. It resembles the structure of the solar system and supplements similar "solar system" models that were proposed by Joseph Larmor and Ernest Rutherford. Whereas *gravitational* forces bind a planet to the Sun, an electron is bound to an atomic nucleus by *electrostatic* forces. In place of the orbital motion of the planets around the Sun, the model envisages negatively charged electrons that orbit the positively charged nucleus of the atom. The angular momentum of an electron that is in orbit around an atomic nucleus is an important physical property defined by the product of three quantities: its mass, the radius of the orbit, and the speed at which it moves around the nucleus. The orbiting electron has an electrical charge, and its motion around the nucleus is equivalent to a tiny current loop, which produces a dipole magnetic field (Fig. 1.4b). The magnetic moment of the dipole is proportional to the angular momentum of the electron's orbital motion. The ratio of the magnetic moment to the angular momentum is called the *gyromagnetic ratio*.

The analogy between the Bohr model of the atom and a planetary system breaks down at the atomic scale. In a planetary system, the radius of a satellite's orbit about a planet is determined by the velocity of the satellite. A continuous spectrum of orbital radii is possible because each one depends only on the amount of energy needed to send the satellite into orbit. However, at the atomic and subatomic scale, the laws of classical physics no longer apply but merge into a different set of laws. At this scale, energy no longer has a continuous spectrum of frequencies but exists in discrete multiples of a fundamental unit called a *quantum*. Quantum theory was introduced in 1900 by the German theoretical physicist Max Planck. It revolutionized the scientific understanding of physical properties at the subatomic scale. As well as energy, other physical quantities such as momentum and angular momentum are quantized.

Particle behavior at the subatomic level is governed by the laws of quantum mechanics, which restrict the possible energy of an orbiting electron to a range of discrete values. As a result, the electrons form distinct concentric shells around the nucleus, each characterized by a different energy level. At the atomic scale, the possible energies are quantized, and the particular shell in which an electron is located is specified by a *principal quantum number*, denoted n. This can have any value from $n = 1$ up to the maximum number of shells in the structure of the atom.

Each shell contains a number of subshells the electron can occupy. They are identified by an *orbital quantum number l*, which is determined by the orbital angular momentum of the electron around the nucleus. The lowest possible energy level within a subshell is called the *s*-orbital and has the orbital quantum number $l = 0$. The next higher-energy level corresponds to the *p*-orbital with $l = 1$; above this lie the *d*-level with $l = 2$ and the *f*-level with $l = 3$ (Fig. 1.6). The ability of an electron to move between different orbitals is restricted by quantum rules, so that not all transitions are possible.

The angle at which an electron orbit is inclined to an arbitrary axis is also regulated. As a result, the angular momentum of an electron and thus also its magnetic moment are quantized. This only affects the electron's energy in a magnetic field. The available energy levels within a subshell are then described by the *magnetic quantum number* m_l, which for each value of l can only have $(2l + 1)$ positive or negative values around l. Thus, for the *s*-subshell, $l = 0$ and $m_l = (2l + 1) = 1$; the *s*-subshell has only one orbital. However, for the *p*-subshell $l = 1$ and $m_l = (2l + 1) = 3$; there are three orbitals corresponding to $m_l = 0$, $m_l = +1$, and $m_l = -1$. The d-level has five different orbitals, with $m_l = 0$, ± 1, and ± 2.

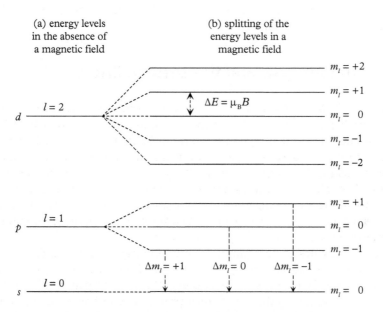

Fig. 1.6 *Illustration of the Zeeman effect. (a) In the Bohr model of atomic structure, the energy levels of an electron in a subshell are specified by different values of the magnetic quantum number l. (b) In the presence of a magnetic field, the dipole magnetic moment due to the electron's orbital angular momentum causes the energy levels to split by an amount ΔE that is proportional to the field.*

Some paradoxical situations are found at the boundary between classical physics and the atomic state. For example, the radius of the electron calculated by classical physics would be larger than the radius of the proton, which is known to have a mass that is 1836 times that of the electron. Moreover, in atomic theory an electron has charge and mass but is considered to be a *point particle*; that is, it has no spatial size. It is difficult to imagine how the self-rotation of a point can have an angular momentum! Nevertheless, analogous to the rotation of a planet about its axis, each orbiting electron has a self-rotation, or *spin*, and angular momentum is associated with it. Consequently, an additional spin quantum number is needed to describe fully the state of the electron.

The rules of quantum mechanics restrict the angular momentum of an electron— and therefore its magnetic moment—to discrete values. This principle applies also to the electron spin because an electron can only spin clockwise or anticlockwise about its axis. So, when two electrons occupy the same orbital around an atom, if one of the spins points "upward," the other must point "downward." This is known as the Pauli exclusion principle, named after Wolfgang Pauli (1900–1958), an Austrian physicist. Consequently, there are only two *spin quantum numbers*, and they have the values

$m_s = +\frac{1}{2}$ or $m_s = -\frac{1}{2}$. In any orbital, there can never be more than two electrons, with opposite spins.

The state of each electron in an atom is uniquely determined by a combination of the four quantum numbers n, l, m_l, and m_s.

1.6   Spectral Analysis and the Zeeman Effect

When energy in the form of heat or light is imparted to an atom, its electrons are elevated to orbitals with higher-energy levels. When they ultimately return to their earlier levels, the electrons emit the extra energy in the form of light. The wavelength of the light is proportional to the energy difference between the two orbitals. However, the orbitals have distinct energies as a result of quantization, and so the light is emitted at specific wavelengths. The light is analyzed with an instrument called a spectroscope, which uses a device called a diffraction grating to separate the wavelengths. These wavelengths appear as a set of distinct lines, each corresponding to a different transition between the internal energy levels of the atom.

Once again, the rules of quantum mechanics impose a restriction (Fig. 1.6). They only allow transitions between energy levels for which the change Δm_l in the magnetic quantum number m_l is +1, 0, or −1 (i.e., $\Delta m_l = 0, \pm 1$). The pattern of electron orbits of a given atom is unique to that atom. As a result, the set of spectral lines for a given atom is produced by electrons moving between a unique set of orbitals and is therefore diagnostic of the atom.

The presence of a magnetic field causes the energy levels of an atom to change. The dipole magnetic moment associated with the orbital angular momentum of an electron interacts with the magnetic field, which changes the energy level of the electron compared to its level in the absence of a field. This phenomenon was discovered in 1896 by a Dutch physicist, Pieter Zeeman, and is named after him. The amount of the energy change, ΔE, in a magnetic field B is given by

$$\Delta E = \mu_B B \tag{1.4}$$

The parameter μ_B in this equation is a fundamental physical constant called the *Bohr magneton*. It is the basic unit of magnetic moment, caused by the angular momentum of the electron's orbital rotation or spin.

The Zeeman effect has wide usage for remotely measuring magnetic fields. In atomic physics, it is a powerful method for studying the internal structure of atoms, whereas in astrophysics it is an important technique for analyzing the magnetic fields of distant cosmic bodies. In particular, the Zeeman effect forms the basis of constructing magnetograms of the Sun, which can be used to construct models of solar magnetic field lines (see Chapter 7.4).

The Sun's magnetic field is enormous in size and extends through the entire solar system, constantly changing in time and influencing the magnetic field of the other planets. Knowledge of the behavior of the Sun's magnetic field helps scientists to understand the internal activity in the Sun. In particular, it helps to explain the emission of electromagnetic radiation from the Sun in the form of waves and particles. These emissions impact the Earth, interact with its magnetic field, and modify the geometry of the field in the near-space environment of the Earth. They can also pose serious threats to life on the planet, especially to the infrastructure on which our modern civilization depends.

1.7 Electromagnetism

Late in the 19th century, James Clerk Maxwell, a Scottish scientist, combined all earlier observations of electrical and magnetic behavior into a set of 20 equations. In modern mathematical notation, they reduce to a set of four vector equations. Maxwell introduced the concept of electrical displacement, which is the effect of an electrical field on charges associated with atoms that are bound to fixed positions. Although the charges remain attached to their atoms, the charge distribution becomes polarized by the electrical field. An alternating field produces an alternating polarization, which results in a changing magnetic field and adds an extra term to the set of equations.

Maxwell's equations collectively describe how an *electromagnetic wave* is formed. It consists of an oscillating electrical field and an oscillating magnetic field at right angles to each other (Fig. 1.7); both oscillations are oriented transverse to the direction of propagation of the wave. Maxwell realized that the family of electromagnetic waves included visible light, and his equations showed that in a vacuum all electromagnetic waves travel at the speed of light.

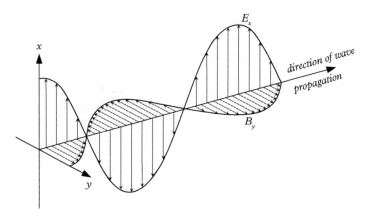

Fig. 1.7 *An electromagnetic wave is composed of an alternating magnetic field at right angles to an alternating electrical field, both of which are perpendicular to the direction of propagation of the wave.*

Michael Faraday demonstrated that a changing magnetic field induced an electrical current in a conducting material. Thus, when an electromagnetic wave encounters an electrical conductor, the alternating magnetic component of the wave induces an alternating electrical current in the conductor. The near-surface free electrons absorb energy from the wave, so that the energy of the wave diffuses away in a short distance below the surface. The decrease in amplitude from an initial value A_0 at the surface to a value A at a depth z is exponential, governed by an equation of the form

$$A = A_0 e^{-z/\delta} \tag{1.5}$$

where the parameter e is known as Euler's constant and is the base of natural logarithms ($e \approx 2.718$). The quantity δ in the equation is called the *skin depth*. It is the depth in which the signal decreases to $1/e$ (i.e., 37%) of its initial value at the surface and is thus a measure of the ability of an electromagnetic wave to penetrate into the conductor. At depths greater than five times the skin depth, only a small percentage of the original amplitude of an electromagnetic wave remains. The skin depth δ depends on the frequency f of the wave and the electrical conductivity σ of the conductor. Its value (in meters) is given approximately by

$$\delta \approx \frac{503}{\sqrt{\mu_r f \sigma}} \tag{1.6}$$

The quantity μ_r is the *relative permeability* of the material. Permeability is a measure of how easy it is to produce a magnetic field in a material. The relative permeability is the ratio of the permeability of a material to that of free space or a vacuum. Commonly, $\mu_r \approx 1$, but in some metals like iron it can be very large, exceeding 10^4. Equation (1.6) shows that the higher the frequency of the electromagnetic wave and/or the conductivity of the conductor, the shallower is the skin depth. High-frequency electromagnetic waves are unable to penetrate deeply into a conductor and are restricted to depths near its surface, whereas lower frequencies can penetrate more deeply into the interior of the conductor.

The study of natural and artificially induced electromagnetic signals in the Earth provides an important tool for locating commercially valuable metallic minerals. Moreover, with increasing depth in the Earth, the temperature rises, which promotes an increase in electrical conductivity. Electromagnetic signals that originate in the magnetosphere and ionosphere induce currents in the body of the Earth, which in turn produce magnetic fields that geophysicists measure and analyze to obtain important information about the planet's internal structure.

1.8 Particle Radiation

The electromagnetic energy radiated by the Sun spans a wide range of wavelengths (Fig. 1.8). At the short-wavelength end there are high-frequency gamma rays that have wavelengths measured in picometers (one picometer is a millionth of a millionth of a meter; 10^{-12} m); at the long-wavelength end there are low-frequency radio waves with wavelengths measured in kilometers. Even lower electromagnetic frequencies are associated with the slow secular variations of the Earth's magnetic field.

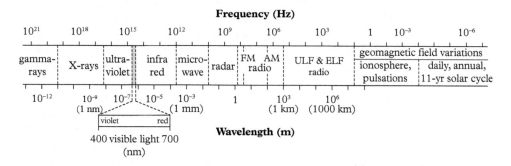

Fig. 1.8 *The spectrum of electromagnetic radiation ranges from very high-frequency gamma radiation to ultra-low-frequency radio waves and very slow changes in the geomagnetic field.*

Visible light represents only a narrow band of the frequencies in the electromagnetic spectrum, corresponding to wavelengths from about 400 nm to 700 nm (one nanometer, abbreviated nm, is a billionth of a meter; 10^{-9} m). The band of visible light is bounded by invisible radiation consisting of long-wavelength infrared radiation at one end and short-wavelength ultraviolet radiation at the other.

The long-wavelength solar radiation that reaches the Earth makes life possible on the planet. The infrared radiation warms the Earth and makes the land and oceans habitable. In conjunction with carbon dioxide and water, some of the energy in visible light is converted at the Earth's surface to sugars and other chemicals by photosynthesis in terrestrial vegetation as well as in phytoplankton in the oceans. Photosynthesis is responsible for the production of oxygen in the Earth's atmosphere. It is an essential source of energy for the existence of life on Earth and has been a key factor in the evolution of life on the planet.

On the other hand, ultraviolet light from the Sun interacts with oxygen atoms in the lower stratosphere at altitudes of 15–35 km above the Earth to form a thin layer that is enriched in ozone. This gas is composed of three oxygen atoms and has the chemical formula O_3; it is less stable than the more abundant diatomic oxygen O_2. The absorption of up to 99% of incident ultraviolet light in the ozone layer prevents much of it from reaching the Earth's surface where it would have detrimental effects on living things and the ecology. Although the concentration of ozone is low, it suffices to shield human life against ultraviolet radiation that would increase the incidence of potentially dangerous skin cancers.

The interaction of solar radiation with the Earth's magnetic field and with the atoms and molecules in the upper atmosphere is quite complex. As explained earlier, classical physics reaches a limit in explaining the behavior of matter at a very small scale, and this is the case for the interaction of solar radiation with the Earth's magnetic field. We can look on solar radiation as an electromagnetic wave of energy, but when we investigate interactions between elementary particles—such as the solar wind with atoms and molecules of the atmosphere—we encounter the paradox of *wave-particle duality*.

The energy in a quantum of radiation is directly proportional to its frequency. Albert Einstein showed that light consists of quantized packets of electromagnetic energy called *photons*. The wave-particle duality provides alternative ways of regarding electromagnetic energy. It can be treated as a classical wave as portrayed in Figure 1.7, or it can be regarded as a stream of energetic particles. This is a conundrum in modern physics, whose fundamental meaning is not without controversy.

The reason why quantum mechanics is consistent with the real world is still not fully understood. For practical purposes, the larger a particle is, the more closely it conforms to classical physics. It is only when elementary particles interact with each other and with the macroscopic world that quantum mechanics becomes necessary to fully explain their behavior. Although the wave-particle duality is puzzling, the predictions of quantum mechanics at the atomic level are borne out by experiment.

2

How the Geomagnetic Field is Measured

Introduction

Shortly after my 22nd birthday, I was employed by a mining company in northern Canada to carry out a geophysical survey during the short sub-Arctic summer. Soon after the snows had melted and the ice had thawed, I found myself in a wilderness of lakes and forests, about 1,000 km north of Winnipeg. As a young immigrant from southern Scotland, where the hills are virtually denuded of old trees, I was a complete "greenhorn" in the primeval Canadian forest. Our camp was close to the tree line, north of which few trees grow. The surrounding forest consisted mainly of bushes and stunted conifers, but their growth was dense enough that it was easy to become disoriented within a short distance of the camp.

That summer I was the scientific leader of a small geophysical exploration team. The other half-dozen members of the team were from the local indigenous Cree tribe, experienced in the woods and expert axe-men, whose main job was to cut parallel paths through the forest undergrowth. Along these prepared tracks my job was to measure the geomagnetic field with a vertical-field magnetometer. At premeasured intervals of 5 to 10 meters, I set up a tripod, mounted the magnetometer on it, leveled and aligned the instrument, and measured the vertical magnetic field. The goal of our survey was to locate magnetic anomalies—deviations from the mean local field—that are caused by magnetic minerals in the veins and dikes intruded into the ancient rocks. An anomaly would indicate a possible concentration of nickel, the metal of greatest interest to the mining company, and warrant exploratory drilling.

The survey was an exciting and interesting experience with a steep learning curve. During the short summer our team successfully located several dikes with weak nickel and copper enrichments. However, to our disappointment, the nickel content of the orebodies was not high enough to be commercially viable, and so they were not developed. My boss explained to me that most projects to find valuable metallic ores suffer the same fate.

Magnetic methods of exploration are inexpensive and deliver important results. As a result, a magnetic survey is often the initial surveying technique used in the exploration for mineral assets in an unknown territory. Local enrichments of magnetic ores in crustal

The Earth's Magnetic Field. William Lowrie, Oxford University Press. © William Lowrie (2023).
DOI: 10.1093/oso/9780192862679.003.0002

rocks may be located directly. Although sedimentary rocks are only weakly magnetic, a magnetic survey can also describe the size and boundaries of a sedimentary basin by measuring the depth to the more strongly crystalline basement rocks that lie beneath the sediments. Magnetic surveys are carried out on land, at sea, from the air, and from satellites. They provide the main source of information needed to describe the Earth's magnetic field on local, regional, and global scales.

2.1   Measurement of Magnetic Field Direction

The technology required to measure the Earth's magnetic field evolved very slowly for many centuries after the invention of early compasses. Mariners realized that navigation by compass resulted in errors in their location because the compass needle did not indicate true north. They called the angular difference between magnetic north and true north the *magnetic variation*; the term is still used in aeronautical charts, but in scientific usage it is called the angle of *declination* (Fig. 1.1). Navigators observed that at different locations the declination can be to the east or west of true north. The *inclination*, or dip, of the field could also be measured with a magnetic needle. When mounted on a horizontal axis, the needle can rotate in a vertical plane. The inclination is measured by orienting the device so that the vertical plane corresponds to the magnetic meridian, and the needle then aligns along the field direction. The angle between its northern end and the horizontal is the inclination; downward angles are positive, upward angles negative.

Well into the 20th century, the most sensitive instruments for measuring the direction of the magnetic field were finely balanced magnetized needles. The most dramatic advance in magnetometer design was made during the Second World War with the development of the fluxgate magnetometer, a robust, sensitive electronic instrument initially developed for use in antisubmarine warfare. It measures the magnetic field in a predetermined direction, and it soon replaced the earlier mechanical instruments. Subsequently, this type of magnetometer was developed further, and its ability to detect weak magnetic fields improved progressively. A modern version can measure the geomagnetic field with a sensitivity of about one part in a million. It is one of the main instruments carried on satellite missions to measure the global magnetic field from space.

2.2   Measurement of Magnetic Field Intensity

When a compass needle is displaced from its alignment with a magnetic field and then released, it oscillates back-and-forward about the original field direction. The stronger the field, the higher is the frequency of the oscillations. By timing the oscillations of a needle about the dip angle of the field in the vertical meridian plane, the apparent field intensity can be measured. The oscillation method does not give the absolute value of intensity because the strength (magnetic moment) of the needle is unknown. However, it allows the observation of *differences* in magnetic intensity from place to place. In 1798 famed German naturalist and explorer Alexander von Humboldt used this method to measure the intensity of the magnetic field at several locations in South America. The results showed differences between the local values and a reference value at a fixed location in Peru.

In 1832 Carl Friedrich Gauss devised a method to measure the absolute value of the horizontal component of the field. The method is a two-stage procedure that for the first time related the dimensions of magnetism to the physical units of mass, length, and time. In a first step, the oscillations of a compass needle about the horizontal field direction are measured; this yields the product MH of the magnetic moment of the needle (M) and the horizontal field strength (H). In a second step, the field produced by the same compass needle is used to cause deflections of a different compass about its north–south alignment in the horizontal plane. These deflections give the ratio M/H of the magnetic moment of the compass needle to the horizontal field strength. The product of the two results gives the square of the magnetic moment, M; the ratio of the first result to the second gives H^2, the square of the horizontal field intensity. The total field intensity, F, is calculated by combining the horizontal intensity with the measured angles of declination and inclination. The Gauss method of measuring geomagnetic field intensity remained in use at observatories for more than a hundred years until the middle of the 20th century.

In recognition of Gauss's contributions to geomagnetism, the unit of magnetic intensity was named after him. The *gauss* served as the unit of intensity until 1960. The adoption of the Système Internationale (SI) definitions of physical units replaced the unit of magnetic intensity by the *tesla*, symbolically written T. This is a very large field, equivalent to 10,000 gauss. It is comparable to the magnetic field at the poles of a permanent magnet made of neodymium or the field produced by a powerful electromagnet. By contrast, the field of the Earth is quite weak; its strength is of the order of 0.5 gauss. However, this measure is much larger than the daily fluctuations in field intensity, or the anomalous fields caused by mineralized rocks in the Earth's crust, which are the target of magnetic exploration. Accordingly, for many years a smaller, practical unit called a *gamma* was used, equivalent to 10 millionths of a gauss (10^{-5} G, or 10 μG). These older units were mostly replaced by modern units, although they are still used in some descriptions of satellite missions (e.g., the successful Juno exploration of Jupiter's magnetic field). The modern practical unit used for expressing geomagnetic intensity is the *nanotesla*, which is one billionth of a tesla and is abbreviated nT. Conveniently, it is equivalent in size to the gamma. The intensity of the geomagnetic field at the Earth's surface varies from about 30,000 nT at the magnetic equator to about 60,000 nT at the magnetic poles.

In order to measure the intensity of a magnetic field with a resolution of 1 nT or better, several types of magnetometer have been developed. A magnetometer is a versatile device that is widely used in geophysical research and exploration. It is also popular for the detection of ferrous archeological artifacts as well as in environmental research. It has military applications in the detection of mines and in the development of weapons.

2.3   Vector Magnetometers

There are two main types of magnetometer. A *vector* magnetometer measures the field component in a selected direction, and a *scalar* magnetometer makes an absolute measurement of the total intensity of the field without regard to its direction. Initially, magnetometers were mechanical devices, but they have gradually been replaced by electronic instruments.

An important development in magnetometry was the invention in 1914 of a mechanical instrument by Adolf Schmidt, a German scientist. The Schmidt-type magnetometer was in principle a magnetic balance designed to measure the vertical component of the magnetic field. It was therefore a vector magnetometer, but it did not measure the intensity of the field directly or completely. It measured the difference between a local field and the field at a reference station where it was calibrated, and therefore it is sometimes referred to as a relative magnetometer. The sensor was a horizontal bar magnet, mounted on two knife-edged pivots of quartz that rested on quartz supports; the magnet was free to rotate in the vertical plane. The deflections of a light beam were recorded optically and calibrated in terms of field strength. The magnetometer was sensitive but needed careful handling because of its delicate construction. It was portable but slow and cumbersome in use, as it needed to be leveled on a tripod. Nevertheless, it was a mainstay in magnetic exploration from the 1920s until the end of the Second World War. During that interval Schmidt magnetometers were installed worldwide at geomagnetic observatories. The sensitivity of the instrument, about 10 nT, was high enough to define the reference magnetic field for the Earth's surface during the first half of the 20th century.

During the Second World War, collaboration between scientists of the U.S. government and the research division of Gulf Oil Company led to the development of a nonmechanical, electrically powered magnetometer. It was much more sensitive than the mechanical type and more robust, and it was easily portable. In wartime service it was deployed successfully on aircraft as a submarine detector. After the war this device, known as a fluxgate magnetometer because of its operational principle, came into widespread use in geophysical investigations of the magnetic field, especially where these could be carried out from an aircraft or a ship.

The simplest sensor of a fluxgate magnetometer consists of two thin parallel rods of *mumetal*, which is a highly permeable magnetic alloy. Even in a weak magnetic field mumetal becomes completely magnetized; the condition is called saturation magnetization. A coil is wound in opposite directions about each mumetal rod, and a detector coil is wound about the pair (Fig. 2.1a). An alternating current in the magnetizing coil induces saturation magnetizations in opposite directions in the two rods, which exactly cancel each other so that there is no signal in the detector coil. In the presence of a magnetic field along the axis of the pair, the antiparallel rods reach saturation magnetizations at slightly different times, so that they no longer cancel each other. The imbalance results in a current in the sensor coil that is a measure of the detected magnetic field.

In an alternative configuration of the device, a driver coil is wound about a circular core of mumetal encased in a sensor coil (Fig. 2.1b). The driving current induces a

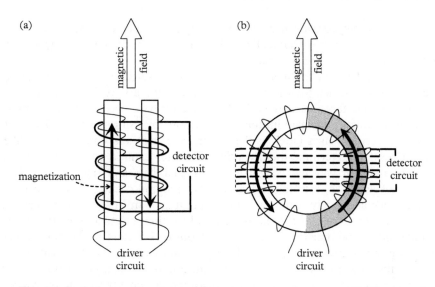

Fig. 2.1 *Basic design of two types of fluxgate magnetometer sensor with (a) a core of parallel linear strips and (b) a ring-type core.*

magnetization that has opposing directions on either side of the symmetry axis, which is the direction along which the field is measured. The sensor circuit detects and measures the imbalance current in an external field, so that the principle of operation of the circular fluxgate is the same as for the straight mumetal rods.

The fluxgate instrument is an example of a *vector* field magnetometer. It measures the strength of the magnetic field in a particular direction, along the axis of the device. However, as a result of its operational design, it must be calibrated in order to convert the output current to magnetic units. The fluxgate magnetometer in effect measures variations in the field relative to a reference value. In some usages, such as making continual recordings of variations in the field at an observatory, it is referred to as a magnetic *variometer*.

A single fluxgate sensor can be used to measure any of the individual magnetic field components. When the axis of the sensor is aligned with the total field, the intensity can be measured. The resolution of a fluxgate magnetometer is on the order of 0.1 nT. Commonly, the instrument consists of three fluxgates at right angles to each other. In some devices the triaxial sensor is rotated automatically until two of the axes give no signal; in this orientation these axes are at right angles to the field, so that the third axis is aligned along the field and can measure its intensity.

A triaxial sensor can also be oriented so that it measures each of the north, east, and vertical field components independently (Fig. 1.1). This has the disadvantage that the instrument must be oriented accurately. However, triaxial fluxgate vector magnetometers are commonly used in satellite missions dedicated to measuring the global geomagnetic field. For example, the Juno mission that investigated the magnetic field of

Jupiter in 2016 employed two triaxial fluxgate magnetometers, each with a resolution of ~ 0.05 nT in its most sensitive range. Each magnetometer made 64 vector measurements of Jupiter's magnetic field per second.

Accurate orientation of the magnetometer aboard a satellite is achieved with a star camera, which is a device for tracking the positions of stars. It therefore operates on an ancient principle that mariners and other travelers have used since time immemorial. They were able to navigate in uncharted regions by using their knowledge of the positions of individual stars (such as the North Pole star) and constellations (like the Southern Cross) to determine their heading. The star camera on board a spacecraft functions in a similar way by recognizing individual stars and star patterns. During centuries of observation, astronomers have measured and catalogued the positions of many stars very accurately. The star camera compares an image of the stars with a database of star positions, and dedicated algorithms then calculate the attitude of the spacecraft (and magnetometer). Inventories of thousands of stars can be stored aboard a spacecraft for this purpose, or they may be consulted by communication with a ground station. On the Juno mission to Jupiter, the attitude of the spacecraft was measured four times per second by comparing star camera measurements with an on-board star catalog.

2.4   Scalar Magnetometers

Magnetometers designed to measure only the total intensity of the field and not its direction are classified as *scalar* magnetometers. An important example is the proton precession magnetometer, which functions by making use of nuclear magnetic properties. Just as the electrons orbiting an atom have a magnetic moment, so also do the protons in the nucleus. This property is used in the design of a proton precession magnetometer (Fig. 2.2). The instrument consists of a flask containing a proton-rich fluid, such as kerosene (or even water), surrounded by a coil that is set preferably at a large angle to the geomagnetic field. A direct current through the coil produces a strong magnetic field, which aligns the proton magnetic moments. When the magnetizing field is switched off, the protons try to realign with the Earth's magnetic field. Analogous to the way that a child's spinning top wobbles about the vertical, so the spinning proton wobbles about the geomagnetic field direction. The wobbling motion is called Larmor precession. Its frequency is proportional to the strength of the geomagnetic field and to the proton gyromagnetic ratio, which is an accurately known physical constant. The precession induces an alternating current in a pick-up coil, which is amplified and displayed directly as the strength of the magnetic field. The proton precession magnetometer can make an absolute measurement of the field with a sensitivity of around 0.1 nT.

The sensitivity of the proton precession magnetometer is exceeded by the Overhauser magnetometer, which operates on the same principle. It achieves a higher sensitivity by using a source fluid enriched with free radicals. These are atoms or ions that have an orbital electron that is unpaired; that is, the electron is the sole occupant of its orbit. The constant exciting field is replaced by a radio-frequency field. It aligns the spins of the

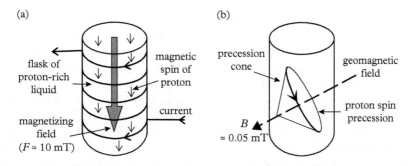

Fig. 2.2 *Principle of operation of a proton precession magnetometer. (a) An electrical current in a magnetizing coil wound around a flask of proton-rich fluid produces a strong magnetic field that aligns the spin magnetic moments of the protons. (b) After the magnetizing field is switched off, the spin magnetic moments precess about the direction of the geomagnetic field.*

unpaired electrons, which then couple to the protons in the source fluid. The electron–proton coupling enhances the signal, which can also be measured faster than the on-off measurement of the proton precession magnetometer. An Overhauser magnetometer has a field resolution of around 0.01 nT.

Another sensitive instrument used in field operations such as archeological and geophysical surveys is the optically pumped alkali vapor magnetometer. It is also used in medical research because it can measure weak magnetic fields from the heart and brain. The principle of the instrument is quite complicated. Like the proton precession magnetometer, it is also based on atomic magnetic properties. Alkali elements such as cesium, potassium, and rubidium have only a single-valence electron (i.e., an electron in the outermost atomic orbital). Light energy, fed into a vapor of the alkali element, excites the atoms and raises their energy state; the process is called optical pumping. As in the principle of the proton precession magnetometer, the electrons precess about the magnetic field at a Larmor frequency that is proportional to the field and to the gyromagnetic ratio of a precessing electron, which is 660 times larger than that of the proton. The frequency can therefore be measured more easily and more accurately. The sensitivity of an optically pumped alkali vapor magnetometer is around 0.01 nT, similar to that of the Overhauser magnetometer.

An absolute field magnetometer aboard the Swarm satellites uses a different principle of operation that gives similar performance. It is an optically pumped magnetometer that uses helium vapor as the source. In accordance with the Bohr model of the atom, the electrons around the helium nucleus can occupy a number of energy levels. Transitions between these levels result in the emission of precise amounts of energy. Atomic spectroscopy shows the transitions as distinct spectral lines (see Chapter 1.6), which are diagnostic for an element. In a magnetic field, the lines split into several components in compliance with the Zeeman effect. The energy differences between the sublevels are proportional to the strength of the magnetic field, which can thus be determined. This method enables measurements of the geomagnetic field with an accuracy of better than 0.3 nT.

An important factor in the interpretation of magnetic anomalies in the Earth's crust is knowledge of the magnetic properties of the rocks that cause them. A special type of rock magnetometer with extremely high sensitivity is needed to measure the weak magnetic signal associated with a rock. In order to function optimally, the instrument needs to be installed in a stable environment that is free of magnetic materials and disturbances, and in which the ambient magnetic field is excluded by special shielding. The most sensitive rock magnetometers are designed around SQUID sensors (the acronym stands for Superconducting QUantum Interference Device), which combine the property of superconductivity with quantum field theory. A SQUID sensor operates at very low temperature, at which the material that forms the sensor becomes superconducting.

A superconductor does not conform to the laws of classical physics. It is a material that, below a critical temperature, has perfect electrical conductivity. The critical temperature is very low, far below ambient temperatures. In this cold state, an electrical current behaves unusually. The absence of resistance means that an electrical current can flow indefinitely; it repels magnetic fields, which cannot enter the superconductor. One of the odd effects is harnessed in geophysics to measure the weak magnetic fields of the Earth as a whole, or, in particular, those associated with crustal rocks and minerals. Known as the Josephson effect, it applies to a situation in which a superconducting circuit is interrupted by a barrier, called a Josephson junction, which may be a thin insulating layer or a narrowing of the conductor. In classical physics the current is blocked. However, at quantum scale, the behavior of individual electrons changes, and the laws of classical physics are replaced by the laws of quantum mechanics. Electrons are able to penetrate the barrier, a process called quantum tunneling, which is assisted by the presence of a magnetic field. The magnetic field at the Josephson junction is quantized; that is, it is made up of individual units of magnetic flux. The unit (or quantum) of magnetic flux is a physical constant determined by the electron charge and the Planck constant, both of which are fundamental parameters; thus, a quantum of flux has the same value in any superconductor.

Classical physics is only able to explain macrosopic behavior and cannot explain this phenomenon. In order to understand the behavior of elementary particles at the scale of individual electrons, atoms, and subatomic particles, it is necessary to delve into the mysterious world of quantum mechanics. That is an advanced topic of modern physics and is far beyond the scope of this book. However, a simple analogy with crowd behavior can convey an impression of how a SQUID sensor functions in a superconducting circuit.

Consider a crowd of people moving along a street in which a barrier has been erected. At the barrier the crowd is halted; nevertheless, some individuals try to cross the barrier and succeed in doing so. If a short ladder is placed against the barrier, it helps more people to cross. Their number depends on how high the ladder is; by counting heads of those who cross, a measure of the ladder's effectiveness, and thus its height, may be obtained. In the analogy we have substituted a superconducting electrical current for the crowd, a Josephson junction for the barrier, electrons for the individuals, and a magnetic field for the ladder. The SQUID sensor functions by counting the number of magnetic flux quanta that pass the Josephson junction, and because the value of a flux quantum is known, the strength of the field can be calculated.

Initially, SQUID sensors were metallic alloys, and the operating temperature had to be lower than the boiling point of liquid helium (4.2 K). However, subsequent research into material properties led to the development of ceramics that become superconducting at higher temperatures, above 90 K. High-temperature (HT) SQUIDs are cooled further and maintained at the temperature of liquid nitrogen (77 K) to create a stable operating environment. SQUID magnetometers have become indispensable in geophysics and geology for measuring the extremely weak magnetizations of rocks and minerals.

2.5 Magnetic Gradiometers

A common method of magnetometry, widely used in mineral exploration, environmental research, and archeology, uses pairs of magnetometers in an arrangement called a *gradiometer*. Two identical instruments are fixed to a rigid mount at a known distance apart, most commonly in the vertical direction. The technology may employ vector magnetometers (e.g., fluxgate) or scalar magnetometers (e.g., proton precession, optically pumped cesium, or Overhauser). It is used to measure the vertical (or horizontal) rate of change of the total field or one of its components. The arrangement places one of the magnetometers closer than the other instrument to a subsurface magnetic source, so that the signals received by the two instruments have different strengths. The difference between the measured fields divided by the separation of the magnetometers is the field *gradient*.

Time-dependent variations of the field such as the diurnal variation are recorded simultaneously by both magnetometers in a gradiometer and do not affect the difference in the signals. Similarly, the magnetic fields caused by distant sources (including the main field generated in the Earth's core) influence each magnetometer almost equally, and their effects largely cancel each other. As a result, the gradient is, to a large extent, due to the near-surface magnetic structure. Gradiometers have better spatial resolution than single magnetometers for the investigation of near-surface magnetic anomalies. They are commonly used in ground-based airborne and marine magnetic surveying and for locating small magnetic targets. Magnetic gradiometers are very useful in archeological surveys for detecting subsurface historic structures, such as buried pits, ditches, and walls that have slightly different magnetic properties than the soils or sediments surrounding them. They are used in environmental geophysics to locate buried ferrous objects, such as historic artifacts. A magnetic gradiometer is an invaluable detector for locating antipersonnel mines in both current and former war zones.

2.6 Terrestrial Magnetic Surveying

A magnetic survey is designed to measure the anomalous crustal fields that arise when rocks with contrasting magnetic properties adjoin each other. The survey may involve measurements from an aircraft, from a ship, or on the ground (Fig. 2.3). For each measurement platform, several adjustments to the measured data are needed in order to interpret the results. These include compensation for the main geomagnetic field with

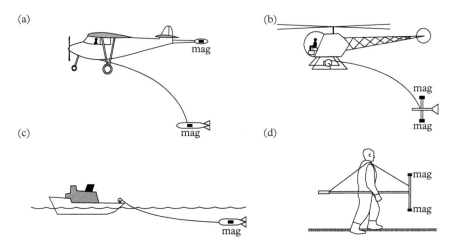

Fig. 2.3 *Sketch of some methods of magnetic surveying: (a) airborne survey with a fixed-wing aircraft using magnetometers in a tail "stinger" and towed "bird," (b) airborne survey with a helicopter and towed magnetic gradiometer, (c) marine survey with a magnetometer towed in a "fish," and (d) ground-based survey with a hand-held gradiometer.*

the aid of the International Geomagnetic Reference Field (Chapter 3.6), corrections for regional trends in the data, and corrections for time-dependent changes in the field (e.g., daily variations) during the measurement interval.

Airborne Magnetic Surveying

An airborne magnetic survey can be carried out with a fixed-wing aircraft or by helicopter. A scalar magnetometer is most commonly used because this type of instrument has high sensitivity, but it does not require special orientation. Typical instruments are the proton precession, Overhauser, and cesium-vapor magnetometers. For high-resolution surveying close to ground level, the instrument can be configured as a gradiometer. The airplane, by nature of its construction, contains a considerable number of magnetic materials. Its magnetic field is compensated by mounting the magnetometer at the wing tips or by installing it in a special extension (called a stinger) that protrudes behind or in front of the plane. Alternatively, the magnetometer may be towed in an aerodynamic housing at the end of a cable on the order of 50 m in length. The location of the sensor relative to ground coordinates is tracked as precisely as possible by GPS navigation.

Airborne surveying can investigate magnetic properties of large areas of the Earth's surface in a short time. This makes the method well suited for regional reconnaissance, as well as for mapping the field on a national scale. It can be employed over inaccessible terrestrial and oceanic terrains, and it is usually the first method used in exploring the subsurface geology of an unknown area. An aerial survey of a region is usually carried out according to a planned flight pattern (Fig. 2.4a). Measurements are made continuously along parallel lines that are oriented normal to the trend of known geological structures. The flight lines are separated from each other by a distance that is chosen to fit the goals of the survey. The closer the line spacing and the lower the flight altitude, the greater is the ability of the survey to resolve small features.

However, the flight altitude must be adapted to the ruggedness of the terrain (Fig. 2.4b). In the reconnaissance of an area with a flat or gently undulating topography, or in a large-scale national survey, the measurement lines may be flown at a constant altitude above sea level using a fixed-wing aircraft. This is not practical over rugged terrain. In this case the survey is flown at a constant height above the topography. This method is viable over rugged terrain with a helicopter, which is capable of flying more slowly than a fixed-wing aircraft. A helicopter is also used when a detailed local investigation is required.

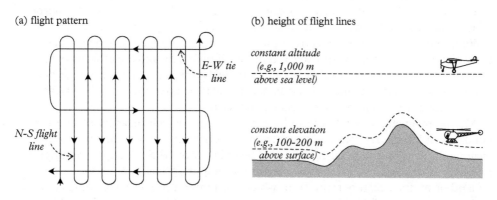

Fig. 2.4 *(a) Common pattern of parallel flight lines and orthogonal tie lines in an airborne magnetic survey. (b) Flight lines at constant altitude above sea level (top) and constant elevation above ground surface (bottom).*

In a typical case, a high-resolution aeromagnetic (HRAM) survey over smooth terrain might be flown at a constant height of 150 m above ground level, with measurement lines spaced 300–500 m apart. A set of tie lines is customarily flown at right angles to the first set, perhaps spaced at the same separation, or up to 3 to 5 times more widely than the main lines. The intersections of the main lines and the tie lines should give the same anomalous field value, so that the cumulative differences around a rectangular loop between adjacent lines should sum to zero. However, small residual differences are usually observed, which arise from errors in location of the aircraft, disturbances due to flight conditions, and the daily (diurnal) variation of the geomagnetic field. Computer algorithms are used to distribute corrections to all the measurements so as to reduce or eliminate these differences.

Much of the world's surface has been covered by airborne magnetic surveys. The largest exceptions are parts of Antarctica and some oceanic areas far from land. Not all of the data are generally accessible because they are often the confidential property of private exploration companies. However, many countries have carried out airborne magnetic surveys of their territories and have released their data for general use.

Marine Magnetic Surveying

The world's oceans cover 71% of the planet's surface. Large tracts of this vast area remain underexplored or completely unexplored scientifically. Satellite measurements, with their near-global coverage, now play a major role in describing the gravity and magnetic fields over oceanic areas. However, the resolution achieved by repeated surveying from a low Earth orbit at 300–600 km altitude above sea level is not as high as can be attained in marine surveying at the ocean surface, only a few kilometers above the magnetic basement. Ship-based surveys are essential for acquiring detailed information about oceanic magnetic anomalies.

In a marine magnetic survey, a magnetometer installed in a streamlined waterproof casing (the "fish") is towed behind a survey ship (Fig. 2.3c) at a distance that is large enough to minimize interference from the large amount of highly magnetic steel in the vessel. The length of the towing cable is typically 200–300 m, and the cable has sufficient buoyancy to maintain the "fish" at about 15 m below the water surface. It is difficult to orient a towed vector magnetometer; hence, a scalar magnetometer of the Overhauser or alkali-vapor total field types is usually employed.

Marine magnetic anomalies have long wavelengths, which measure several tens to hundreds of kilometers, but early surveys were confronted by the difficulty of exactly locating the ship's position far from land. The ubiquity of GPS has largely removed this problem. A detailed marine survey may use a temporary base station located nearby on land or on the ocean bottom, from which the position of the magnetometer can be tracked. Corrections for daily variations in the external field are also needed, which may be made by reference to a suitable base station or observatory on land. Where the reference station is too distant, cross-tie lines can be used, as is customary in an aeromagnetic survey. Marine anomalies can have large amplitudes, ranging up to more than 1,000 nT, much larger than the corrections for diurnal field variations except on days of high solar activity.

The tracks of marine geophysical ships still do not cover the oceans uniformly. There are large gaps in coverage of oceans in the southern hemisphere and in the Pacific Ocean. This is partly due to the large area and inaccessibility of these oceanic regions, but it is also a consequence of the slowness of marine surveying. For reasons of efficiency and cost minimization, marine magnetic surveying is usually carried out in conjunction with other geophysical measurements, such as gravity and seismic profiling, which by their nature require the vessel to sail at slow speed.

Archeological Magnetic Surveying

In addition to providing important information for evaluating a country's mineral resources, magnetic surveying plays a useful role in investigating its ancient culture.

Earlier civilizations have left evidence of their existence in artifacts and structures that in due course became buried, along with metallic artifacts, perhaps including iron or steel weapons or implements. Ancient burial sites may produce magnetization contrasts, for example, between buried walls and surrounding soils. Entire buried villages and towns have been mapped with magnetic surveying, especially using a magnetic gradiometer. The buried objects and local ground disturbances at ancient burial sites may cause small magnetic anomalies of only a few nanotesla, but such tiny anomalies can be measured well with modern magnetometers carried by an operator (Fig. 2.3d), especially in a gradiometer configuration. A detailed small-scale magnetic survey of an inferred archeological site can provide useful evidence for deciding upon its potential value and subsequent excavation.

2.7 Magnetic Observatories

The geomagnetic field varies irregularly over the surface of the Earth and also changes with time, over both short and long intervals. In order to describe the spatial variation of the magnetic field and to analyze it mathematically, measurements are required from sites located all over the planet. Long-term series of measurements are required for analysis of time-dependent changes in the field. In the 1830s Carl Friedrich Gauss and Wilhelm Weber set up a network of measurement stations, called the *Magnetischer Verein* (i.e., magnetic society or association), at which the geomagnetic elements were measured systematically. At first, magnetic observatories were located primarily in Europe. Alexander Von Humboldt promoted the setting up of permanent measurement stations for making long-term records of the magnetic field elements. He successfully involved the British and Russian Empires as participants, and thus the first global network of magnetic observatories was established. It comprised 53 observatories at which frequent measurements of the magnetic field were made at regular intervals. Although the accuracy of the early data is low by comparison with modern standards, the continuity of measurements covers a very long time and the quality has been constantly upgraded. The early records can still be incorporated in modern analyses of historic, large-scale changes in the main magnetic field.

The original observatory network was sparse, with large regions of the Earth underrepresented. The number of magnetic observatories has increased with time, so that presently there are more than 180 of them (Fig. 2.5). To obtain long-term records, an observatory must occupy a stable fixed location, which restricts its location to a land area, that is, a continent or island. However, the global distribution of usable land areas is uneven. On the one hand, oceans cover three-quarters of the Earth's surface and islands are not uniformly scattered in them; on the other hand, many regions on land are inaccessible wilderness. As a result, the worldwide distribution of observatories is still spatially uneven and located predominantly in the northern hemisphere.

The requirements that must be fulfilled to qualify as a geomagnetic observatory are very stringent. They encompass the location and construction of the building,

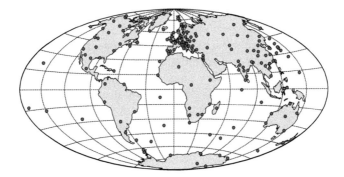

Fig. 2.5 *The global distribution of active geomagnetic observatories. (*Data source: *International Association for Geomagnetism and Aeronomy, Division V)*

instrumentation, and measurement procedures. The location must be in an area of low magnetic noise, that is, free from local magnetic disturbances; for this reason, magnetic observatories are remote from population centers and traffic. The local geology must also be taken into account: the rocks should not cause magnetic gradients that would produce a large change in the field across the site. Typically, the observatory is housed in a building kept at a stable temperature and is constructed from nonmagnetic materials by excluding iron and other ferromagnetic constructional elements. Ideally, an individual small building is used to house each magnetometer, mounted on a stable pillar. The instrumentation includes both vector and scalar magnetometers with a resolution of 0.01–0.1 nT. In the early 19th century, the observatory measurements were made visually and recorded manually at prescribed intervals (e.g., 60 minutes), but it was both difficult and expensive to maintain this schedule. In 1847 the Royal Observatory at Greenwich, England, introduced a method of recording hourly observations photographically, which also provided more continuous records of temporal field variations. Subsequently, these observations were recorded on magnetic tape. At modern observatories, the measurements are recorded digitally at a much higher acquisition rate, at least every minute and often every second, and they are shared internationally within hours.

Magnetic observatories have historically provided two important types of information: accurate measurements of the magnetic field and its variation with time over many years. Until the start of the space age in the early 1960s, the long sequences of continual observations at magnetic observatories were the principal sources of data for constructing the International Geomagnetic Reference Field (IGRF). This is a model of the global field at the Earth's surface, compiled by international cooperation between many countries (Chapter 3.6). It is the reference field with which magnetic surveys of the Earth's crust are compared in order to define anomalous regions.

In 1987 a consortium of observatories, the International Real-time Magnetic Observatory Network (INTERMAGNET), was formed to share data and to establish modern standards of operation. The network currently consists of more than 150 observatories, maintained by more than 40 countries. A member observatory uses a vector fluxgate magnetometer to measure field components and a scalar proton precession (or Overhauser) magnetometer to measure the absolute field intensity. The continual variometer data are averaged at minute intervals and sent to a Geomagnetic Information Node (GIN)—one of five observatories that function as collection points for data from the INTERMAGNET observatories. The one-minute average field values are distributed from the nodes to the rest of the network by satellite and internet, thus supplying the INTERMAGNET observatories with real-time data. The high-capacity digital recording has led to an increase in the number of observatories capable of providing continuous vector data at one-second intervals and scalar data at 30-second intervals. As a result, the one-minute data are gradually being supplanted by one-second data.

Prior to the space age in the early 1960s, long-term observations at magnetic observatories provided the main source of data for constructing maps of the global field. An abundant supply of high-quality geomagnetic measurements now comes from dedicated satellites in low Earth orbit. The satellite data have a global coverage superior to that of the observatories. Not only do satellite data provide the components of the present field, but they now represent a time interval long enough to observe the time-dependent secular variations of the field.

The data from magnetic observatories complement the satellite data and continue to be important for longer-term monitoring of changes in the field. The observatories deliver continual, highly accurate field measurements at a particular location over time intervals lasting from seconds to decades. These temporal changes are important for understanding the magnetic field that surrounds the Earth—the magnetosphere. The external interactions of the field with Earth's space environment are summarized in numerical indices that are computed from observatory data. The indices are used to monitor various phenomena that constitute "space weather" and the threats they pose for the Earth (Chapter 8.5).

2.8 Satellite Mapping of the Global Magnetic Field

More than 5,000 scientific satellites were in orbit around the Earth as of 2022. Their orbits can be grouped in three categories—geosynchronous, medium orbit, and low Earth orbit—according to their orbital radius (Fig. 2.6). The first three paragraphs in this chapter describe each of these categories in turn. A large number have circular orbits that coincide with the equatorial plane. At a radial distance of 42,164 km (altitude 35,786 km), these satellites circle the Earth at exactly the same rate as the Earth's daily rotation about its axis. Consequently, the orbits are *geosynchronous*, with a period of exactly one sidereal day (i.e., the time for a 360° rotation relative to a fixed star, 23 h 56 m 4 s). Relative to a fixed position on the Earth, the satellites appear to be stationary in the sky. As a result, the orbits are also said to be *geostationary*. This property allows them to be used by television stations, for weather satellites, and for communications between spacecraft and control stations on Earth. At the end of its operational

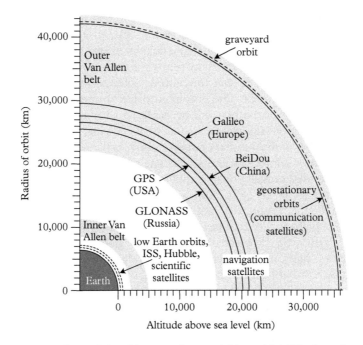

Fig. 2.6 *Some of the orbits presently occupied by artificial Earth satellites. The Hubble telescope, International Space Station (ISS), and most scientific satellites occupy low-Earth orbits beneath the Inner Van Allen radiation belt (see Chapter 8.2). Navigational satellites (e.g., GPS) orbit at about 20,000 km altitude and circle the Earth approximately twice daily. Geostationary communication satellites are in an equatorial orbit at a distance of about 6.6 radii from the center of the Earth. Navigational and geostationary orbits lie within the Outer Van Allen radiation belt.*

life a geosynchronous satellite may be moved into a slightly higher "graveyard orbit" to reduce the possibility of collisions with other satellites.

Closer to the Earth's surface, at altitudes of approximately 19,000–23,000 km, satellite orbits are classified as medium Earth orbits. These satellites provide services such as navigation and positioning for civilian and military purposes. They are known familiarly as the American GPS, Russian GLONASS, European Galileo, and Chinese BeiDou systems. Each system consists of 18–50 satellites, whose orbits are distributed in several planes inclined to each other. The navigational satellites have orbital periods of roughly 12 hours and are linked in networks that provide global coverage and accurate navigation.

The orbits at altitudes of approximately 400–600 km are classified as low Earth orbits (LEO). There are almost 2,000 satellites in LEO orbits, engaged in various scientific and commercial missions; in addition, there are thousands of small, mass-produced satellites of the Starlink internet constellation. The LEO orbits are high enough that drag from the residual atmosphere at that altitude is sufficiently low to allow scientific satellites to operate for years and even decades, yet they are close enough to the Earth to allow good resolution of scientific observations. The best known of the LEO satellites are the Hubble Space Telescope and the manned International Space Station, the latter of which has been operational since the year 2000. A number of LEO satellites are designed to make precise measurements of physical properties, for example, of the planetary gravitational and magnetic fields. However, only a few satellites have been primarily dedicated to measuring the geomagnetic field.

A geomagnetic satellite typically orbits the entire Earth about 14–15 times per day while the Earth spins beneath it, so the global coverage of satellite measurements in just a single day is more complete than the network of observatories. In the course of a multiyear mission, satellite magnetic data provide near total coverage of the Earth's surface. However, satellite orbits are located in the F_2-layer of the ionosphere (see Fig. 2.3), so the measurements are subject to changing ionospheric magnetic fields and must be carefully selected or corrected in order to minimize these signals. Together, the data from satellites and observatories complement each other and allow separation of contributions to the measured field from different sources.

Systematic surveys of the magnetic field from low Earth orbit were initiated in 1964. Three of the six Polar Orbiting Geophysical Observatory (*POGO*) series of satellites carried scalar magnetometers, and from 1964 to 1971 they measured the intensity of the magnetic field from steeply inclined near-polar orbits. After a gap of 14 years, the next dedicated LEO geomagnetic survey—the *MAGSAT* mission (1979–1980)—was launched in a near-polar orbit at 325–550 km altitude. In addition to a cesium-vapor scalar magnetometer, the satellite carried a triaxial fluxgate magnetometer with which it made the first global vector survey of the magnetic field.

Twenty years later, in 2000, the Danish *Ørsted* satellite was launched in a near-polar low Earth orbit, at an altitude of 650–850 km. The Ørsted satellite was equipped with an Overhauser scalar magnetometer with accuracy 0.5 nT and a fluxgate vector magnetometer with resolution 0.1 nT. Attitude errors limited the accuracy of the vector magnetometer to 2–8 nT. The Danish mission was followed by the German *CHAMP* mission (2000–2010), which had a steeper orbit of 87° and orbited at a lower altitude of 250–450 km. It deployed a scalar magnetometer with accuracy 0.5 nT and a vector magnetometer with accuracy 2 nT. Both missions were operational for periods that greatly extended the planned surveys.

The superior ground coverage obtained by a satellite is illustrated by the ground track of a single day of the Ørsted satellite (Fig. 2.7). The inclination of the near-polar orbit was 97°, which confined the track between latitudes 83°N and 83°S, leaving an unsampled polar gap with radius 7°. Starting at 57°S 72°E, the satellite moves northward (*heavy solid line*), crossing the equator at 58°E (*solid arrowhead*). After traversing the polar region, it moves southward (*heavy dashed line*) and crosses the equator at 134°W (*solid arrowhead*). The time between the equatorial crossings is 50 minutes, one-half of the orbital period. After an additional 50 min, the northward-moving satellite again crosses the equator at 33°E (*open arrowhead*), 24° westward of the first crossing. The ground coverage of a single day of the satellite's track is clearly superior to that of the

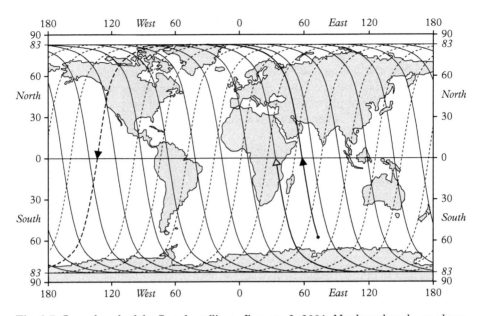

Fig. 2.7 *Ground track of the Ørsted satellite on January 2, 2001. Northward tracks are shown by solid lines, southward tracks by dashed lines. Arrowheads mark successive equatorial crossings of the satellite. (*Redrawn after *N. Olsen, G. Hulot, and T. J. Sabaka, Sources of the geomagnetic field and the modern data that enable their investigation.* Handbook of Geomathematics, *W. Freeden et al. (eds.), Springer, Berlin, Heidelberg, 2015. With permission from Springer Nature)*

irregularly, often widely, spaced ground observatories (Fig. 2.5) The satellite orbit in the later CHAMP mission had a steeper inclination of 87°, giving a smaller polar gap of 3° and more closely spaced ground coverage.

The most recent LEO mission, *Swarm*, launched in 2013 and still operational as of 2022, has the specific goal of measuring the internal field in such detail that the part produced by magnetized rocks in the Earth's crust and lithosphere can be accurately resolved. An original configuration of three identical satellites was employed for the mission. As of 2022, two of them (known as Swarm A and C) fly almost side by side at an altitude of around 450 km, and the third (Swarm B) is in a higher orbit at around 510 km altitude. Each satellite carries latest-generation scalar and vector magnetometers. The satellite orbits are inclined at 87.35–87.75° to the equator, and thus they nearly pass over the poles of the Earth's rotation axis. The gravitational attraction of the Earth on the orbital plane of the lower pair is slightly different from its attraction on the upper satellite because of their different altitudes. The difference causes the orbits to drift at different rates. This means that the planes of the orbits shift slowly relative to the Earth and to each other. In 2018 the higher orbit of Swarm B became perpendicular to the lower orbit of Swarm A and C. In 2021 the higher orbit had drifted so far that Swarm B orbited the Earth in the opposite sense to the lower satellites. During this phase, the flight controllers were also able to reduce the along-track separation of Swarm A and C to a mere 2 seconds, to obtain higher-resolution magnetic measurements. The orbit of Swarm B is now again drifting apart from the lower orbit.

The abundance of data from LEO satellite missions in the past 60 years has made it possible to map the Earth's magnetic field in ever finer detail. The coefficients of the International Geomagnetic Reference Field and their secular variations have been computed with greater precision. Numerical models of the field and its time changes have been developed to ever higher harmonic degrees. In conjunction with long-term data from magnetic observatories, the satellite data have contributed to a more detailed picture of the global magnetic field at the surface and a better understanding of the processes in the Earth's core that generate it.

2.9   The Geomagnetic Field at the Earth's Surface

The direction and intensity of the field vary from place to place at the Earth's surface. The locations on a map, where the field has the same value, may be connected by contour lines, similar to the way altitudes are marked on a topographic map. When the measured value of the magnetic field is different from the value expected from a model, or by comparison with other measurements, the difference is called a magnetic *anomaly*. Contour lines are commonly used to map anomalies, outlining places that differ by the same amount from a reference value. An alternative method of mapping the field is to use a multicolored plot in which different colors are associated with mapped values or anomalies. This method results in a smooth map with continuous changes of color and tone, which gives an immediate overview of the anomalous field.

The contour lines of equal inclination (Fig. 2.8) show that the inclination is 90°—that is, the magnetic field dips vertically downward—in the north of Canada, at a location referred to as the northern *dip pole*. For the geomagnetic reference field in 2020, this pole is located at 86.5°N, 164.0°E. The dip is vertically upward in a region between Australia and Antarctica, which defines the southern dip pole (at 64.0°S, 135.9°E). Note that these locations do not lie directly opposite to each other geographically. The magnetic dip equator, where the field is horizontal, projects as a sinuous line across the map.

Mathematical analysis of the field (Chapter 3) shows that its strongest component corresponds to a dipole located at the center of the Earth. The places where the axis of this geocentric dipole intersects the surface are called the *geomagnetic poles*. They lie symmetrically opposite each other. Their positions in 2020 were at 80.6°N 72.7°W and 80.6°S 107.3°E, so that the dipole axis was tilted at 9.4° to the rotation axis. Recall that the geomagnetic poles and the dip poles have different locations because the geomagnetic poles are defined for only the geocentric dipole part of the magnetic field, whereas the dip poles are defined for the total magnetic field. The locations of the dip poles and geomagnetic poles change slowly with time by a small amount each year due to secular variation.

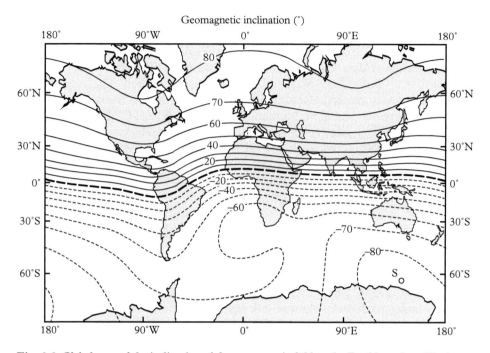

Fig. 2.8 *Global map of the inclination of the geomagnetic field at the Earth's surface. The heavy dashed line is the magnetic dip equator, where the field is horizontal. The letter S marks the location of the southern dip pole, where the field is vertically upward; the northern dip pole is located off the top of the map. (Redrawn after E. Thébault, C. C. Finlay, et al., International Geomagnetic Reference Field: the 12th generation. Earth, Planets and Space, 2015. Creative Commons License)*

The intensity of the Earth's magnetic field ranges from more than 60,000 nT in high northern and southern latitudes to less than 24,000 nT in equatorial latitudes (Fig. 2.9). Its maximum values in the northern hemisphere are close to—but not coincident with— the northern poles; there is also a high value in northern Russia. The maximum intensity in the southern hemisphere is greater than 66,000 nT south of Australia, over the southern dip pole. However, the minimum intensity is not located on the magnetic equator, as would be expected for a purely dipole field. The lowest value covers a large region that includes parts of South America and the South Atlantic Ocean, and extends almost to South Africa. Here the geomagnetic field is unusually weak, measuring less than 24,000 nT at the Earth's surface instead of more than 30,000 nT elsewhere at a comparable magnetic latitude.

The feature is called the South Atlantic Anomaly. It results from the processes by which the geomagnetic field is generated in the Earth's liquid outer core. The low field intensity is caused by the flow pattern of the core fluid, which produces reversely polarized magnetic flux at the surface of the core under this region of the southern hemisphere. The flow pattern in the core is not constant but changes slowly with time, with

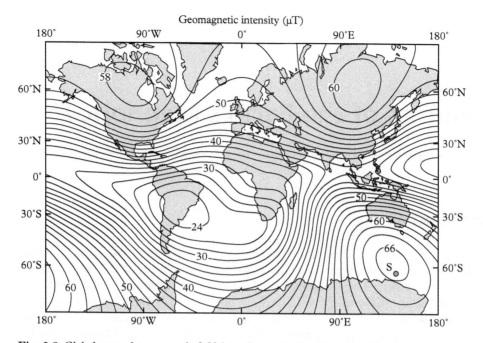

Fig. 2.9 *Global map of geomagnetic field intensity at the Earth's surface (1 μT = 1,000 nT).* (Redrawn after E. Thébault, C. C. Finlay, et al., International Geomagnetic Reference Field: The 12th generation, Earth, Planets and Space, 2015. Creative Commons License)

accompanying effects on the surface magnetic field. The analysis of magnetic data from the Swarm satellite mission has revealed that the South Atlantic anomaly is changing relatively quickly for a geophysical process. During only 6 years of the mission, from 2014 to 2020, the anomaly increased in area and extended itself further eastward toward southern Africa.

3

Sources of the Earth's Magnetic Field

Introduction

Although magnetism was known to Chinese civilizations, possibly as early as 5,000 years ago, they had no evidence concerning its origin, and as a result they regarded it as a supernatural power. They believed that the north–south orientation of a compass needle must be due to the attraction of a heavenly object, such as the Pole star or a prominent asterism like the Big Dipper. As a result, early compasses were used to align buildings and structures so that they conformed to the practices of *feng shui*, a Chinese form of geomancy. Roman philosophers attributed a metaphysical origin to magnetism. Although magnetic compasses were used for navigation in medieval times, the force that oriented them remained a mystery. Eventually, the work on magnetized spheres of lodestone by Gilbert and le Nautonier at the beginning of the 17th century showed that compass needles pointed north because the Earth's magnetic field is a property of the planet. However, none of the observations cast a light on where the sources of the field were located within the Earth. The existence of postulated magnetic mountains or islands in the far north was believed widely enough in the late 16th century that Gerardus Mercator, the Dutch cartographer, showed a 53-km-wide mountain (designated the Rupes Nigra, or Black Rock) near the north pole of his early maps.

The 17th-century Age of Enlightenment opened the way to a realistic understanding of how the geomagnetic field is produced. In the first half of the 19th century, mathematical analysis of careful measurements of the field allowed Carl Friedrich Gauss to establish beyond doubt that the main field arises from sources deep in the Earth. Gauss, of course, could not know that the Earth has a liquid core, which was only discovered by seismologists early in the 20th century, and is where the main part of the field is generated. Subsequently, numerous precise observations of the field have been acquired, initially only at magnetic observatories but now predominantly from geomagnetic satellite missions. Mathematical processing of the data with modern computers allows further separation and identification of the field's different sources. A small fraction of it is generated in the upper reaches of the Earth's atmosphere, and part is due to ancient fields recorded in the rocks of the Earth's crust. The planet's internal structure, together with the temperature and pressure at great depths, are important factors that create the conditions for producing the geomagnetic field.

The Earth's Magnetic Field. William Lowrie, Oxford University Press. © William Lowrie (2023).
DOI: 10.1093/oso/9780192862679.003.0003

3.1 The Earth's Internal Structure

A century of recording and analyzing the passage of seismic waves through the Earth's interior has enabled seismologists to develop a detailed model of the planet's internal structure. An earthquake generates vibrations that propagate away from its source, or *focus*, and are recorded by extremely sensitive instruments, *seismometers*, at the surface; the record is called a seismogram. The place on the surface of the Earth vertically above the focus is called the *epicenter*. The great-circle distance along the surface between the epicenter and a seismometer is called the epicentral distance. It is usually measured in degrees of arc.

The energy released in an earthquake travels through the interior of the Earth as so-called body waves. When the seismic energy reaches the Earth's surface, it spreads out along it in the form of surface waves. Each category of waves is further divided into two types, characterized by the kind of particle motion they embody. The body waves are subdivided into compressional waves (called P-waves) and shear waves (S-waves). In a P-wave the ground vibrates backward and forward along the direction of propagation, causing a succession of compressions and expansions of the ground through which it passes. An S-wave travels by means of a shearing motion, in which the ground shakes at right angles to the direction of propagation. It can only travel through a material that supports shear, and so it cannot travel through a liquid. The P-wave is the fastest body wave and is therefore the first seismic signal to arrive at a detector on the surface. The S-wave travels through solids at only about 70% of the P-wave velocity, but it is still faster than the surface waves.

The two types of surface waves are called Rayleigh waves and Love waves, respectively. Their energy is, by definition, restricted to motions of the near-surface layers of the Earth. In a Rayleigh wave, the surface moves backward and forward at the same time as it moves up and down. This produces an elliptical motion of the surface, similar to the way a floating cork bobs on the surface of the sea. The Love wave is a horizontal shear vibration that exists only when there is a low-velocity layer at the surface.

When a seismic body wave encounters the interface between two layers with different seismic velocities, part of its energy is reflected and the rest crosses the interface. The transmitted wave experiences a change of direction at the interface, known as refraction. The seismic waves from an earthquake bounce around the Earth's interior and are reflected and refracted at each interface they encounter. The seismogram of an earthquake therefore contains multiple arrivals that are superposed upon each other. Seismologists have developed sophisticated ways of identifying the arrivals of different waves on the seismograms. As a result, body waves—including their multiple refractions and reflections—travel through the Earth at speeds that are well known at every depth.

The travel times of the seismic waves allow subdivision of the interior of the Earth into a layered structure of concentric shells. To a first approximation, the layered structure resembles a 4-minute egg: a thin hard *crust* (the brittle egg shell) surrounds a firm *mantle* (the solidified egg white), inside of which there is a liquid *core* (the runny yolk). However, in contrast to an egg, the Earth's liquid core surrounds a small, solid inner core. Seismology has established many fine details of the crust–mantle–core structure, which help us understand how the geomagnetic field is produced.

Seismic P-waves and S-waves that travel through the crust, mantle, and core experience multiple reflections and refractions at the different interfaces (Fig. 3.1). The outermost shell of the Earth is the hard, brittle crust. It varies in thickness from 5–10 km beneath the oceans to as much as 60 km under mountain ranges on the continents. The shell beneath the crust, called the mantle, extends to a depth of 2,891 km, where it abuts the top of the core. The uppermost layer of the mantle is also brittle and together with the crust forms a hard outer layer called the *lithosphere*. It has variable thicknesses, extending to depths of 70–100 km under the oceans and 100–150 km under the continents. The lithosphere blankets the surface of the Earth and is divided into a number of thin plates. Although a few minor plates are no larger than a European country, each of the 10 major plates extends horizontally for several thousand kilometers. Internal dynamic processes in the Earth keep the plates in constant motion, causing earthquakes where they are in contact with adjacent plates.

An approximately 150-km-thick layer of the upper mantle immediately beneath the lithosphere has lower rigidity than the lithosphere and is called the *asthenosphere* (which means that it is "weak"). The upper and lower boundaries of the layer are not sharply defined. At depths up to a few hundred kilometers, the mantle reacts rigidly to stress and is the source of deep earthquakes, which occur where the stress overcomes the local

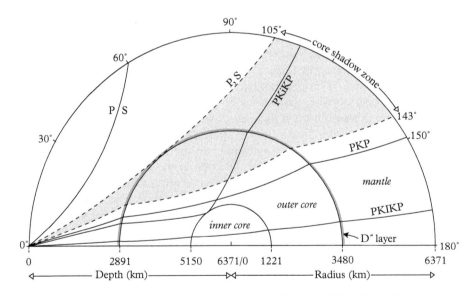

Fig. 3.1 *The paths through the Earth of seismic P- and S-waves. Direct P- and S-waves from an earthquake return to the surface within an epicentral distance of 105°, beyond which no direct S-waves are observed. The P-wave that traverses the outer core, refracted as it enters and leaves, is labeled PKP; the P-wave through the inner core is labeled PKIKP; and the P-wave reflected at the surface of the inner core is labeled PKiKP.*

strength, particularly at the margins of tectonic plates. At greater depths the mantle has a dual character. It reacts like a rigid solid on short time scales (e.g., during the passage of a seismic wave), but over long time intervals it flows slowly like a very viscous fluid.

Seismic travel times have identified a thin layer, about 200 km thick, at the bottom of the mantle that has unusual physical properties. Based on an earlier alphabetic nomenclature for identifying the different layers of the Earth's interior, it is called the D″ layer (or D-double-prime layer). It plays an important role in mantle dynamics and may influence the flow of heat out of the core. The thickness of D″ varies laterally, in a similar way to the Earth's crust, and forms structures on the mantle–core interface that resemble continents. D″ is believed to be a source of plumes of hot magma that rise through the mantle and erupt at the surface as hotspots, which are an important feature of global plate tectonics.

The paths of seismic waves change dramatically at a depth of 2891 km (Fig. 3.1), which marks the boundary between the solid mantle and the liquid core beneath it. The existence of a liquid core was postulated in 1906 by Richard Oldham, a British scientist, who noted the delayed travel times of seismic P-waves that passed through the central region of the Earth. In 1914 Beno Gutenberg, a German seismologist, confirmed the existence of the core and calculated its size. Seismic P-waves and S-waves that skim the surface of the core return to the Earth's surface at an epicentral distance of 105° from an earthquake. No direct S-waves are observed at epicentral distances beyond 105°, indicating that they do not pass through the core. S-waves can only travel in a rigid material, and their absence indicates that the outer core is liquid. It consists of iron in a liquid state because the temperature in the outer core is well above its melting point.

Seismic P-waves travel through the liquid core with reduced velocities because a liquid allows the passage of rhythmic compressions but cannot support shear. They are refracted twice in transit: the direction of travel changes abruptly when they enter the core and again when they leave it. After traversing the liquid core, the P-waves return to the surface at epicentral distances beyond 143°. No direct P-waves arrive at epicentral distances between 105° and 143°, which bound a seismic "shadow zone."

The shadow zone is not completely free of P-waves, however. Some weak arrivals are observed in the gap between 105° and 143°. They are interpreted as P-waves that have passed through the liquid outer core and were reflected at the surface of a solid inner core. This evidence for a solid inner core was first recognized in 1936 by a Danish seismologist, Inge Lehmann. For many years her interpretation was controversial, but it was finally corroborated by superior data from seismic arrays. The solid inner core forms because the increasing pressure causes the melting point of iron to increase more rapidly with depth than does the actual temperature. Eventually, the pressure becomes so great that, despite the high temperature of more than 6,000 degrees, the molten iron alloy solidifies to form the inner core. As a result, the inner core is growing in size at the expense of the outer core. The radius of the inner core is 1,221 km, which means that it is smaller than the Moon and similar in size to the dwarf planet Pluto.

The inner core is becoming progressively better understood as geophysical data accumulate. Seismic observations, in particular, have revealed a structure that is more complicated than described above. Once regarded as a uniform sphere of solid iron, the inner core is now known to have a layered structure and to be anisotropic to the passage of seismic P-waves. Their velocity is estimated to be about 1% faster in an axial direction than in the equatorial plane.

In addition to establishing the internal structure, seismic data provide depth profiles of the Earth's density and elastic parameters. The fact that all magnetic fields must originate in electrical currents requires that the core be an adequate electrical conductor. Geochemical and cosmochemical arguments constrain the possible composition, which is understood to be broadly similar to that of iron meteorites. The core is estimated to consist largely of iron (~ 85–88%) alloyed with nickel (~ 5%) and smaller amounts (3–6%) of lighter elements like oxygen or silicon. Trace amounts of many other elements are also present, but opinion is divided as to whether the core contains radioactive elements such as potassium or uranium. The geochemical model of core composition does not require them, but some geophysicists argue that a radioactive source may be necessary to provide the energy to power the geodynamo that produces the geomagnetic field.

3.2   Pressure and Temperature in the Earth

Pressure and temperature both increase with depth in the Earth (Fig. 3.2). The physical unit in which pressure is measured is called a pascal (Pa). It is the pressure when a force of one newton acts on a surface of one square meter. In this unit the standard atmospheric pressure at mean sea level is equal to 101,325 Pa. Inside the Earth the enormous

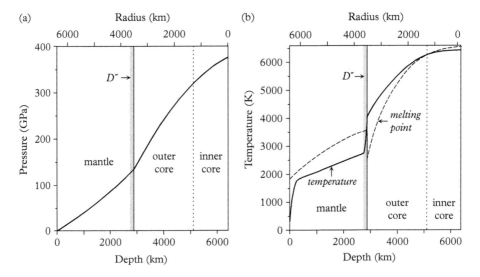

Fig. 3.2 *Estimated radial profiles of (a) pressure and (b) temperature and melting point in the Earth. (After Figs. 7.37 and 9.2 in W. Lowrie and A. Fichtner,* Fundamentals of Geophysics, *3rd ed., Cambridge University Press, 2020. Reprinted with permission)*

pressure is measured in *gigapascal* (GPa), equivalent to a billion pascal (10^9 Pa). The pressure at a given depth is determined by the cumulative weight of the overlying layers of the Earth on each unit of area. This causes the pressure to increase continuously and smoothly with increasing depth. There is a notable increase in slope at the core–mantle boundary, due to the sharp increase in density in the transition from the silicate rocks of the lower mantle to the liquid iron in the outer core (Fig. 3.2a). The pressure at the center of the Earth is estimated to be 380 GPa, which is almost 3.8 million times the atmospheric pressure at sea level.

Temperature is commonly measured on the Celsius (°C) or Fahrenheit (°F) scale. With a few exceptions, the latter scale is used only in the United States and its territories, while the Celsius scale has widespread usage in the rest of the world. Each scale has advantages and disadvantages in daily usage. Absolute zero, the temperature that is often thought of as the lowest temperature possible, is defined by international agreement to be −273.15 °C (−459.67 °F). In scientific research the unit of temperature is called a kelvin (K). In size it is equivalent to a degree Celsius (°C), but the units are displaced so that 0 K is equated with −273 °C. A temperature of 0 °C is therefore equal to 273 K.

At shallow depths below the Earth's surface, the temperature increases rapidly at an average rate of around 20–30 K/km. However, the rate of increase with depth—the temperature gradient—is much lower in the mantle than in the crust, and it is estimated to increase by only about 0.3 K/km in the lower mantle. The mantle temperature measures about 2,500 K near to the core–mantle boundary (CMB). Above this level it remains well below the melting temperature of the silicate minerals that form the mantle (Fig. 3.2b), and as a result the mantle behaves like a viscoelastic solid. This means that it responds like a rigid solid to sharp, brief disturbances, such as the forces that accompany the passage of a seismic wave, but on a geological time scale of millions of years it flows like a very viscous liquid. The flow of a solid material at high pressure and temperature is due to defects in the crystal structure of a mineral, such as vacancies and dislocations. At high temperatures in the mantle they become thermally activated and migrate through the solid minerals. Over geological time intervals, the motion of crystal defects results in a slow displacement of mantle material, enabling it to flow at an estimated speed of a few cm/yr, similar to the rates of motions of lithospheric plates at the surface.

The boundary between the core and the mantle represents a dramatic change in composition and is accompanied by a sharp increase of temperature with depth. The temperature climbs from 2,500 K to 4,000 K across this important boundary. The melting point of iron in the outer core below the CMB is about 2,600 K, which is well below the temperature of 4,000 K at that depth. As a result, the iron in the outer core is molten and the iron is able to flow. How easily it can flow is determined by its *viscosity*. This is a common property of everyday materials; for example, water is less viscous and flows more easily than a "sticky" syrup.

To understand what causes viscosity, imagine a horizontal layer of fluid that is flowing with a certain velocity above a slower layer of fluid. Some molecules from the upper layer diffuse into the slower layer; this transfers their momentum to the lower layer, speeding it up. Similarly, molecules from the slower layer transfer upward into the faster moving layer, slowing it down. The exchange of molecules between adjacent layers in the flow transfers momentum and has the effect of producing cohesive forces between the layers, which cause the fluid's "stickiness" or *viscosity*.

The viscosity of molten iron at the temperature and pressure in the outer core is estimated to be similar to that of molten iron at the Earth's surface, and it flows with comparative ease. The estimated rate of flow of the core fluid is on the order of several millimeters per second (mm/s), which amounts to tens of *kilometers* per year. This is much faster than the rate at which continuous geological motions take place at the Earth's surface. For example, seafloor spreading causes the lithospheric plates to move apart horizontally at divergent plate boundaries on oceanic ridges at only a few tens of *millimeters* per year. Flow in the Earth's mantle takes place at a rate that is comparable to global plate motions.

3.3   Dipole and Multipole Fields

Magnetic fields are classified as either toroidal or poloidal, on the basis of the geometry of their field lines (Fig. 3.3a). A toroid is a closed surface with a hole in the middle; that is, it is shaped like a doughnut. A circular solenoid, consisting of an iron ring with electrical wires wound continuously around it, illustrates this type of magnetic field (Fig. 3.3b). Each winding of the wire forms a loop that creates a magnetic field, and the overall field of all the windings is confined to the interior of the ring. Toroidal magnetic fields are

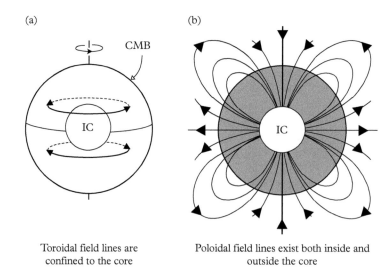

Toroidal field lines are
confined to the core

Poloidal field lines exist both inside and
outside the core

Fig. 3.3 *(a) Toroidal magnetic fields form closed loops within the Earth's core and cannot be measured outside it. (b) Poloidal fields, as in this sketch of a quadrupole, exist within the Earth and can be measured at and above its surface.*

employed in plasma physics in the design of a *tokamak*. This device is used to confine extremely hot plasmas for research in thermonuclear fusion. Toroidal magnetic fields are present in the interiors of the Sun and stars, where the field lines form ring-shaped belts around the rotation axis. Although they cannot be measured at the Earth's surface, toroidal fields play an important role in the outer core in generating poloidal fields.

A poloidal magnetic field has closed field lines that begin and end on fictitious "poles," as in the case of a bar magnet or William Gilbert's lodestone. The field lines are not confined to the source region and are detectable outside it. Thus, the poloidal fields generated in the core characterize the geomagnetic field outside the Earth. The poloidal concept is useful for describing the geometry of magnetic fields and lends itself to a powerful mathematical technique for analyzing them.

The simplest magnetic field is that of a dipole (Fig. 3.3), which consists of a single pair of oppositely signed poles that are so close together that they are effectively at the same point. This is the most fundamental type of magnetic field. It is symmetric about an axis, and its intensity varies with distance r from the "center" of the dipole and also with azimuthal angle θ relative to the dipole axis. In geographic coordinates the angle θ is called the magnetic co-latitude. The field of a magnetic dipole of strength m (which is called its magnetic moment) is a vector, **B**. It has a radial component B_r and an azimuthal component B_θ, given by the equations

$$B_r = \frac{\mu_0}{4\pi}\left(\frac{2m\cos\theta}{r^3}\right); \quad B_\theta = \frac{\mu_0}{4\pi}\left(\frac{m\sin\theta}{r^3}\right) \tag{3.1}$$

The quantity μ_0 is called the *magnetic constant* and is one of the fundamental physical constants. It is equivalent to the permeability of free space and has the value $4\pi \times 10^{-7}\,\mathrm{N/A^2}$.

To obtain the strength B of the dipole field at a distance r and co-latitude θ the expressions (3.1) are summed and squared, which yields

$$B = \frac{\mu_0 m}{4\pi r^3}\sqrt{1 + 3\cos^2\theta} \tag{3.2}$$

More complicated multipole fields always consist of combinations of an even number of poles, again because magnetic monopoles do not exist. If two dipoles are brought together end-to-end at a point, the constellation is called a quadrupole (Fig. 3.4). It has four (i.e., 2^2) poles; because of the index 2, the quadrupole field is said to be of degree 2. If two quadrupoles are brought together, they form an octupole with eight (i.e., 2^3) poles, and so the octupole field has degree 3. The next higher constellation, combining two octupoles, has 16 (i.e., 2^4) poles and degree 4. Increasingly, more complicated geometries can be generated by doubling the number of poles for each increase in degree. A field of degree n would correspond to 2^n poles. The Earth's magnetic field is very complex, and to express it requires a combination of multipole fields overlaid upon each other, in such a way that the sum of the different parts (dipole, quadrupole, octupole, etc.) equals the observed field.

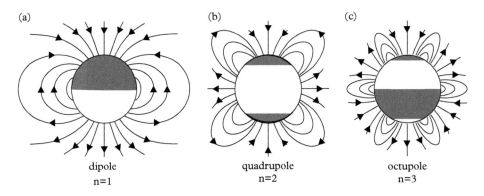

(a)   (b)   (c)

dipole   quadrupole   octupole
n=1   n=2   n=3

Fig. 3.4 *The field-line geometries of (a) dipole, (b) quadrupole, and (c) octupole fields. Each field is rotationally symmetrical about its axis. Shading marks alternating zones where the magnetic field lines leave or enter the surface of a sphere.*

The strength of the magnetic field produced by each multipole diminishes with increasing distance r from it. The field of a dipole (degree $n = 1$) decreases as $1/r^3$; that is, it is inversely proportional to the cube of the distance from the center of the dipole. The field of a quadrupole ($n = 2$) decreases as $1/r^4$; and so on. In general, a multipole field of nth degree decreases with distance from the multipole source as $1/r^{n+2}$. This provides a useful way of calculating the intensity F_r of a multipole field of nth degree on a surface that is at a distance r from the source, if the value F_R is known on a different surface with radius R. The ratio of the fields on the two surfaces is given by

$$\left(\frac{F_r}{F_R}\right)_n = \left(\frac{R}{r}\right)^{n+2} \tag{3.3}$$

When r is less than R, the calculation is called *downward continuation*; when r is greater than R, it is called *upward continuation*. The continuation methods are used to project measurements from one radius to a different radius in general analysis of the magnetic field, as well as in investigations of the anomalous magnetic fields related to geological structures.

Upward continuation of measurements made on a surface near the source of a signal to a surface that is further from it reduces the amount of detail in the result. For example, when data from magnetic surveys that were made at different altitudes are combined by upward continuation to a common surface, a smoother pattern is obtained in which local details may have been lost. By contrast, a disadvantage of downward continuation of measurements from a higher surface to one nearer to the source is that the high-degree part of the signal is amplified relative to the lower degrees. Consequently, noise in the measurements is augmented along with the signal of interest. This consideration has to be taken into account, for example, when geomagnetic data measured at satellite elevation are downward-continued to the surface of the Earth or the core–mantle boundary. The mathematical procedures in upward and downward continuation involve specialized data processing to obviate this problem.

The geomagnetic field varies with both latitude and longitude over a spherical surface around a magnetic source. When multipole fields of different degree are superposed, the resulting appearance depends on the relative importance of the amplitude of each term. It changes with radial distance due to the different spatial dependencies of the individual terms and with direction relative to the source. It is not a simple matter to describe such a field mathematically. To do so, a mathematical method called spherical harmonic analysis is employed. The method was developed in the late 18th century by the French mathematician Pierre-Simon, Marquis de Laplace, to analyze the gravity field on the Earth's spherical surface. Carl Friedrich Gauss later adapted the method to describe the geomagnetic field at the Earth's surface as a sum of terms representing superposed multipole fields.

3.4   Internal and External Sources of the Magnetic Field

When magnetic fields are superposed, it is a difficult task to calculate the resultant field because the individual fields are vectors and their different directions have to be taken into account. Instead of adding the magnetic fields directly, a normalized potential energy (called the potential) is calculated for each field. This is a scalar property, and thus the total potential of the superposed fields can be found by simply adding the individual potentials. The directional components of the resulting field are then calculated by mathematically differentiating the resulting total potential.

To clarify the concept of potential, consider the potential energy of an object in a gravitational field. For example, when a boy climbs a ladder to get a book from a high shelf, his gravitational potential energy is greater than when he was standing on the floor. A book on the same high shelf also has gravitational potential energy relative to the floor, but its mass is less than that of the boy, so its potential energy is less than his. The gravitational *potential* is defined as the gravitational potential energy per unit mass and is calculated by dividing the potential energy by the mass of the object. Thus, both the book and the boy have the same gravitational potential on the high step of the ladder, because it is determined by the difference in their position relative to the floor and not by their different masses. The vertical gradient of the potential is the acceleration of gravity.

In a similar way, a potential can be defined for the energy of the geomagnetic field. It is a measure of how the energy of the field varies with radial distance from the Earth's center as well as with position on a surface at that radius. The gradient of the potential in any direction gives the magnetic field in that direction. In order to describe how the potential varies on the surface of a sphere, special functions called spherical harmonic functions are required. The theory of these functions is too complex to handle in depth here, but it is instructive to look at how geophysicists use them to analyze the Earth's magnetic field. The potential itself is not measured directly. Instead, it is computed by fitting a model of the potential to field measurements. The data for the computation are acquired at a range of altitudes, including ground observations and data from different satellites. The model of the potential can then be used to calculate magnetic field components at any point on other concentric spherical surfaces.

Spherical harmonic analysis of the geomagnetic potential at the Earth's surface is divided into two parts, corresponding to the effects of magnetic field sources that lie, respectively, inside and outside the Earth. Let the potential of the internal sources be V_{int} and that of the external sources be V_{ext}. The total potential V is the sum of these two contributions: $V = V_{int} + V_{ext}$. It was already clear from Gauss's analysis in the 1830s that the potential of the external field is very small in comparison to that of the internal field.

Nevertheless, the external field cannot be neglected because it can affect measurements of the internal field. It originates in the uppermost atmosphere of the Earth and in a "plasma" that occupies space close to the Earth. Plasma is one of four physical states—along with solid, liquid, and gas—in which matter can exist. It consists of a very hot gas of electrically charged particles (i.e., electrons, protons, and ions) that have so much energy that they are not able to combine, and so they coexist as an ionized gas. The motions of these particles constitute electrical currents, which produce magnetic fields that add to the geomagnetic field in space near the Earth.

The source of the external magnetic field's energy is solar radiation (Chapter 8). Part of the external magnetic field is produced in the *ionosphere*. This is an electrically charged part of the upper atmosphere, which is energized by solar heating on the day-side of the Earth and by high-energy particles from the Sun in polar regions. An important component of the external field is caused by the solar wind, a fast-moving plasma emitted by the Sun. Captured by the geomagnetic field, the solar plasma fills the near-space around the Earth and forms the *magnetosphere*. Both sources of the external field, in the ionosphere and magnetosphere, are energized by the Sun, and thus the Earth's rotation causes them to vary with time throughout the day at any given place. The external fields induce electrical currents in the atmosphere as well as in the solid Earth and oceans. The induced currents in turn cause secondary magnetic fields that add to the primary magnetic field of internal origin.

In addition to modulating the regular behavior of the external field, the Sun's activity is subject to sudden changes that affect the external field. Although transient, these solar-driven disturbances can have serious consequences for science and society. In order to account for the disturbance fields, numerical indices are computed at geomagnetic observatories and applied as needed, for example, in selecting and correcting satellite and terrestrial measurements. By reason of their influence on the electrical currents in the magnetosphere and ionosphere, the effects of the Sun on the geomagnetic field are very important. They define the concept of space weather (Chapter 8.5), which is the scientific study of the effects of the Sun's activity on our civilization. Its goal is to predict unfavorable and potentially dangerous solar events and to design suitable countermeasures.

3.5 Spherical Harmonic Analysis of the Internal Field

The internal field originates in the body of the Earth and has two separate but related sources. The main part is the core-generated field, but it also induces magnetizations in the Earth's crust and lithosphere, which contribute their own magnetic fields to the overall field. Any measurement of the field contains both components. They can be distinguished by mathematical analysis of the potential energy of the magnetic field, V_{int}, using a technique called spherical harmonic analysis. This is a powerful method of analyzing the variation of the potential on the surface of a sphere. The method was developed in the 18th century and applied by Gauss to a global set of measurements from the worldwide network of magnetic observatories that he and his colleague, Wilhelm Weber, had set up. A brief look at the principles of the method illustrates how it provides important information about the origins of the internal field.

The potential is defined for any position on a spherical surface concentric with the Earth. Three parameters define a position in spherical geometry. The first parameter is the distance r from the center of the coordinate system (in this case the center of the Earth). A constant value describes a spherical surface. A position on the sphere is defined by its co-latitude θ and by its longitude ϕ. If R is the radius of the Earth, the geomagnetic potential, V_{int}, at a point with coordinates (r, θ, ϕ) is given by the equation

$$V_{int} = R \sum_{n=1}^{\infty} \sum_{m=0}^{n} \left(\frac{R}{r}\right)^{n+1} (g_n^m \cos m\phi + h_n^m \sin m\phi) P_n^m (\cos \theta) \qquad (3.4)$$

This complicated expression represents many terms. It includes a double summation (each indicated by the symbol $\sum$) over all values of n between 1 and a maximum value (which in this case is infinity), and for all values of m between 0 and n. It couples the radial variation of potential with its dependence on latitude and longitude. The function $P_n^m (\cos \theta)$ on the right of the equation is called an associated Legendre polynomial, named after the French mathematician Adrien-Marie Legendre, who discovered this special set of functions in 1782. It describes how the potential varies with the co-latitude θ. The sine and cosine functions of longitude ϕ describe how the potential varies around a circle of co-latitude θ. The product of the variations with co-latitude and longitude describes how the potential varies on a spherical surface and is called a spherical harmonic function. The contributions of different spherical harmonics to the total potential are determined by the magnitudes of the coefficients g_n^m and h_n^m, which are called the Gauss coefficients of the field of degree n and order m. The values of the Gauss coefficients are important parameters for understanding where different parts of the geomagnetic field originate within the Earth.

The spherical harmonic functions have the appearance of sinusoidal waves that undulate around circles of latitude and longitude. The line where a wave intersects the surface of the sphere is called a nodal line. The pattern of nodal lines for a particular combination of n and m defines alternating domains on the surface of the sphere where the undulating surface representing the potential deviates from that of a sphere (Fig. 3.5), alternately being elevated above the sphere or depressed below it. The spherical harmonic function is zero on *m* circles of longitude and (*n—m*) circles of latitude. The functions for which *m* = 0 are called *zonal* harmonics; they form alternating zones parallel to circles of latitude. Functions for which *n* = *m* describe alternating sectors between circles of longitude and are called *sectoral* harmonics. Functions for which *n* and *m* are unequal form a checkered pattern between nodal lines consisting of the *m* circles of longitude and (*n–m*) circles of latitude they are known as *tesseral* harmonics.

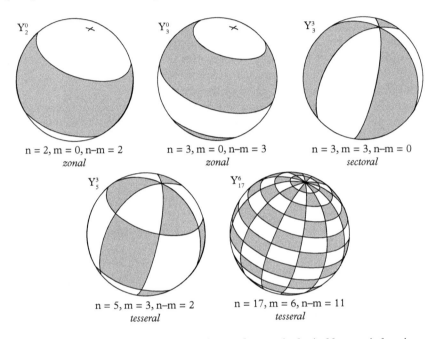

Fig. 3.5 *Examples of zonal, sectoral, and tesseral types of spherical harmonic functions. With increasing degree* n *and order* m, *the number of "patches" increases, enabling the functions to represent ever smaller details on the surface of a sphere.*

The harmonics have different wavelengths. The longest wavelength is the Earth's circumference, while the shortest wavelengths correspond to regional features. Clearly, the higher the degree of an analysis, the more detailed the model of the field. The periodic nature of the sine and cosine functions implies that they are continuous around the Earth. Thus, one can associate a wavelength with each degree in the spherical harmonic analysis. For a model with a large number of degrees, the highest of which is N, an approximate value of the shortest wavelength λ that can be resolved is found by dividing the circumference of the Earth ($2\pi R$) by N. Using 6,371 km for the radius (R) of the Earth, we find that the minimum resolvable wavelength is

$$\lambda = 2\pi R / \sqrt{n(n+1)} \approx 40,000/N \, \text{km, for large } N. \tag{3.5}$$

The relative importance of each wavelength is expressed by the coefficients g_n^m and h_n^m, which represent its amplitude. To analyze a field on a chosen surface, a large number of Gauss coefficients g_n^m and h_n^m must be calculated so that the mathematical model fits the measurements as closely as possible. By summing a large number of wavelengths, weighted according to their relative strengths, a complex field can be described on the spherical surface. The radial dependency of the potential (Eq. 3.4) allows the field to be calculated for a spherical surface at any radial distance from the source. Commonly used surfaces for portraying the field are the altitude of a satellite orbit, sea level, or a chosen depth in the Earth such as the surface of the outer core.

In his 1838 analysis of the global geomagnetic field, Gauss used measurements from a variety of sources and interpolated them at 84 locations. The 24 coefficients of the spherical harmonic functions up to degree 4 were then calculated. Performed without the aid of computers or even mechanical calculators, the analysis must have been a very demanding task, requiring painstaking care. Gauss used the coefficients to calculate the modeled field at the locations where the field was known and found that the calculated values agreed well with the measured values. His results confirmed that the most important part of the geomagnetic field corresponds mathematically to a magnetic dipole located inside the Earth.

In summary, the geomagnetic field is made up of several components (Fig. 3.6). The fields of external origin produce short-term changes that reflect the influence of the Sun. The internal field can be separated into two parts. The main part originates in the fluid core and provides a reference field that can be computed at any place on the globe. A smaller part is induced in the rocks at the Earth's surface, some of which also possess remanent (i.e., permanent) magnetizations. The detailed measurement and analysis of the crustal magnetizations provide geological data that can be exploited for important economic uses. The remanent magnetizations of rocks also provide important contributions to understanding how the core-generated field has changed through geological time.

3.6   The International Geomagnetic Reference Field

At the present time, more than 150 ground-based magnetic observatories worldwide cooperate in a network called INTERMAGNET. They make accurate measurements of the present-day geomagnetic field and provide valuable records of how the field is changing with time over lengthy time intervals. Since the 1970s, observatory data have been augmented by satellite data, which have superior global coverage and comparable high quality, but they cover a shorter history. Modern analyses of the geomagnetic field are based on enormous amounts of data which, with the aid of a supercomputer, make it possible to calculate a large number of Gauss coefficients (hundreds in some models).

The results have produced progressively more accurate and more detailed models of the magnetic field. The International Geomagnetic Reference Field (IGRF) is the standard model of the main geomagnetic field. It results from an international project

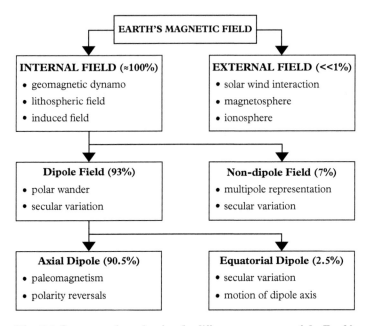

Fig. 3.6 *Summary chart showing the different components of the Earth's magnetic field. The internal field is subdivided here into the dipole component, which has long-term geological significance, and the nondipole component, which is described by higher degree terms in the spherical harmonic analysis. Some of the higher-degree terms originate inside the core and change with time. The terms that describe the lithospheric field are due to induced and remanent magnetizations and are largely unchanging.*

involving the cooperation of geophysicists at many academic and governmental institutions. The IGRF forms the basis for many research studies, such as the generation of the field in the Earth's core field, the definition of local and regional magnetic anomalies related to crustal rocks, and the description of space weather. The reference field is used by commercial companies and governments that want to exploit natural resources, and it is also important for navigation.

From 1900 until 1995, the IGRF was defined for every 5-year interval by the Gauss coefficients of the spherical harmonic analysis up to and including degree and order 10. Since 1995, the high-quality data from satellite missions have made it possible to define the IGRF more precisely to degree and order 13. Many more coefficients of even higher degree are computed from the data, but, as explained below, they have a different origin than the main field. The truncation of the coefficients that define the IGRF at $n = 13$ is deliberate, designed to avoid overlap between the IGRF and the higher-degree part of the field.

The Gauss coefficients of degree $n = 1$ (i.e., g_1^0, g_1^1, and h_1^1) define the strength and direction of a geocentric dipole with three orthogonal components. The strength of the dipole is called its *dipole moment (DM)* and is computed from the Gauss coefficients by

$$DM = \left(\frac{4\pi R^3}{\mu_0}\right) \sqrt{\left(g_1^0\right)^2 + \left(g_1^1\right)^2 + \left(h_1^1\right)^2} \tag{3.6}$$

As before, R is the Earth's radius and μ_0 is the magnetic constant.

The largest Gauss coefficient, g_1^0, represents a dipole aligned with the rotation axis and is referred to as the axial dipole. The coefficients g_1^1, and h_1^1 describe smaller dipole components in the equatorial plane, at right angles to the rotation axis and to each other. The equatorial dipole g_1^1 intersects the equator at the Greenwich meridian, and the dipole h_1^1 intersects it at 90 °E.

Although a geocentric inclined dipole is the dominant feature of the geomagnetic field, an inclined dipole located slightly away from the center of the Earth gives a slightly better mathematical fit to the observations. The location of this *eccentric dipole* can be calculated by including the five *quadrupole* ($n = 2$) coefficients along with the three *dipole* coefficients. This combination defines a best-fitting eccentric dipole centered at 23 °N, 137 °E. The location is displaced from the center of the Earth by about 590 km into the northern and Pacific hemispheres. The offset may indicate an asymmetry in the processes that generate the field.

It is important to remember in the context of analyzing measurements of the field that the multipole representation is a mathematical convenience that allows geophysicists to describe the complexity of the real field. In reality, there are no multipoles located at the center of the Earth. The success of spherical harmonic analysis in describing the field does not imply that separate physical current systems are responsible for each part of the field. In reality, the field is generated by complex currents in the fluid outer core, which create a field that becomes increasingly complex as the source of the field in the core is approached.

The Gauss coefficients vary slowly with time, a phenomenon known as *secular variation*. The most familiar manifestation of secular variation is the need to correct compass needles periodically. The secular variation in declination—also referred to as the *magnetic variation* (e.g., by sailors and aviators)—has been studied since the 17th century (see Chapter 1.2). The predicted rates of change of the Gauss coefficients up to degree and order 8 are computed every 5 years for the next 5-year interval as part of each new IGRF version. These allow corrections to the coefficients during the following 5-year interval for which this version of the IGRF is valid. Magnetic observations for the past four centuries define the historic secular variation of the geomagnetic field. It is conveniently described by separating the behavior of the geocentric dipole from the higher degree, nondipole part.

Repeated determinations of the IGRF for every 5-year interval since 1900 show that the magnetic moment of the dipole has been weakening during the past century (Fig. 3.7a). Expressed in terms of its field at the magnetic equator, the dipole magnetic moment has been decreasing by ~ 2.0 μT per century (1 μT = 1,000 nT). If it continues to weaken at this rate, the intensity will approach zero in about 1,500 years. However, it is not really possible to extrapolate the field behavior so far into the future on the basis of the short length of time for which the IGRF has been calculated. The tilt of the dipole to the rotation axis is around 10° (Fig. 3.7b), but since 1980 the magnetic axis has been slowly creeping closer to the rotation axis at a mean rate of 4.4 degrees per century. Since 1940, the longitude of the geomagnetic pole (Fig. 3.7c) has been moving westward at a mean rate of 5.4 degrees per century, which is equivalent to a few kilometers per year at the Earth's surface. These changes in the field are driven by processes in the core, which are fast compared to the time scales of other geological processes. For example, plate tectonic motions at the Earth's surface take place at rates on the order of a few centimeters per year.

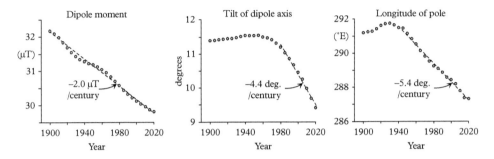

Fig. 3.7 *Historical secular variations of the dipole field during the 20th century, computed from the first-degree coefficients of the IGRF-12 reference field. (a) The dipole magnetic moment, expressed in terms of the strength of the equatorial magnetic field; (b) the tilt of the dipole axis to the rotation axis, and (c) the longitude of the geomagnetic pole.*

The Gauss coefficients of degree higher than one ($n > 1$) describe what is called collectively the *nondipole field* (NDF). Temporal changes in the coefficients cause secular variation of both the NDF and the dipole. A comparison of the maps of the vertical component of the NDF for the years 1780 and 1980 shows the main features of its secular variation (Fig. 3.8). It is characterized by two parts, one of which is a stationary (standing) component and the other a moving component. The maps are dominated by large-scale positive and negative anomalies, which vary in both position and strength. Between 1780 and 1980, positive anomalies over Asia and North America and over the Southern Ocean remain stationary in position but increase in amplitude; the negative anomalies become more strongly negative and those in equatorial latitudes move slowly westward. The motion is referred to as *westward drift*. The changes in the strength and position of the NDF anomalies are caused by changes in the flow pattern of the liquid iron close to the surface of the core.

3.7   Spatial Power Spectrum of the Internal Field

The variation of the magnetic potential on the surface of a sphere is described by the superposition of smooth, periodic functions of position (Eq. (3.4)). The higher the

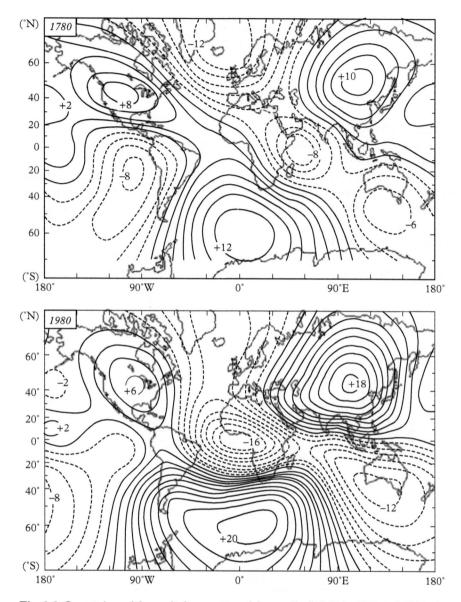

Fig. 3.8 *Comparison of the vertical component of the nondipole field in 1780 and 1980; the contour-line interval is* 2μT. *(After Fig. 11.12 in W. Lowrie and A. Fichtner,* Fundamentals of Geophysics, *3rd ed., Cambridge University Press, 2020. Reprinted with permission)*

degree n of the potential function, the shorter is the wavelength it represents. The energy associated with a particular wavelength of a sine or cosine function is called its *power*. It is proportional to the square of the amplitude of the wavelength. Thus, the power P_n

associated with the terms of degree n in the potential of the magnetic field is computed by summing the squares of the $(2n + 1)$ coefficients of degree n and is given by

$$P_n = (n+1) \sum_{m=0}^{n} \left[(g_n^m)^2 + (h_n^m)^2 \right] \tag{3.7}$$

The Gauss coefficients have the same units as the magnetic field (i.e., nanotesla, nT); thus, the units of magnetic power are nT^2. A plot showing the power at each degree n is called a *spatial power spectrum*.

The MAGSAT mission in 1979 made it possible to calculate accurately a large number of terms in the geomagnetic potential. The spatial power spectrum of the field measured by MAGSAT is shown in Fig. 3.9. Note that the vertical axis of the figure is plotted on a logarithmic scale, on which each successive unit represents an order of magnitude change in power.

The spectrum has several notable features. First, the power associated with the dipole term $(n = 1)$ is much stronger than all other terms, confirming that the geomagnetic field

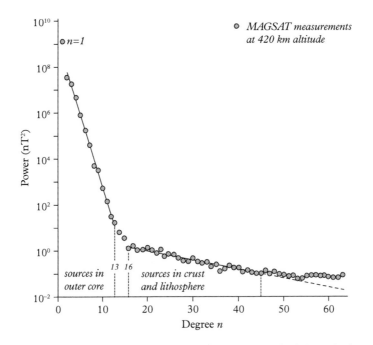

Fig. 3.9 *The spatial power spectrum of the geomagnetic field determined by the MAGSAT satellite at an altitude of 450 km. The part of the spectrum associated with degree n $\leq$ 13 is used to model the International Geomagnetic Reference Field. (Data source: J. C. Cain, Z. Wang, D. R. Schmitz, and J. Meyer, 1989. The geomagnetic spectrum for 1980 and core-crustal separation. Geophysical Journal, 97, 443–447.)*

at and above Earth's surface is dominantly that of a geocentric dipole. The quadrupole term ($n = 2$) is much weaker than the dipole, and the octupole term ($n = 3$) is proportionately weaker than the quadrupole. Higher-order terms become progressively smaller; the power associated with the 13th degree is about 10^4 times less than the power of the dipole.

A second notable feature of the spatial power spectrum is the linear decrease in power with increasing degree for $n \leq 13$. The coefficients with degrees 2 to 13 are fitted well by a straight line on the semilogarithmic plot. The slope of the line can be used to compute the depth of the source of this part of the field. At a particular depth where every degree in the spectrum has equal power, the slope would be horizontal, and the spectrum is then said to be "white." It is the maximum depth of the sources of the terms $n = 1$ to $n = 13$, and it occurs at a distance of about 3,300 km from the center of the Earth. The result indicates that the main part of the internal field originates in the outer part of the Earth's liquid core, which has a radius of 3,480 km.

A further feature of the MAGSAT power spectra is the change in slope of the plot for degrees n larger than 16. The terms from $n = 16$ to $n = 45$ again lie on a well-defined straight line. Above $n = 45$ the MAGSAT data deviate from the line, due to noise introduced by inaccuracies in determining the very small terms of higher degree. Later satellite missions have greatly improved the definition of the higher degree terms that compose this part of the spectrum. The slope of the straight line indicates that the source of this part of the internal field is in the Earth's lithosphere, that is, in the hard outer shell of solid minerals that makes up the crust and uppermost part of the mantle. The minerals become magnetized by the Earth's magnetic field at the present time, and some minerals, in addition, possess a permanent magnetization, which they acquired during their formation.

The part of the geomagnetic power spectrum for degrees 14 and 15 cannot be attributed to a single source because the core field and the crustal field overlap in this part of the spectrum. For this reason, the International Geomagnetic Reference Field, which describes the main core field, is truncated at $n = 13$. The lithospheric part of the magnetic field has a constraining effect on the ability of space missions to investigate higher degree details of the field generated in the deep interior of the Earth.

In the MAGSAT mission, which took place almost a half-century ago, the Gauss coefficients of the IGRF were computed with a resolution of 1 nT. Subsequent dedicated magnetic field missions—such as Ørsted, CHAMP, and Swarm—have yielded more precise measurements of the Gauss coefficients, with a resolution of better than 0.1 nT. The high-degree part of the spatial power spectrum has been extended to much higher degrees than in the MAGSAT model. For example, the CHAMP mission lasted 10 years and measured the Gauss coefficients up to degree and order 100, which allows accurate mapping of large-scale magnetic anomalies with wavelengths greater than 400 km. The goal of the current Swarm mission, launched in 2013, was to determine high degree terms in the magnetic spectrum even more accurately, to $n = 150$ and wavelength 260 km. The data define the magnetic elements more exactly and allow the preparation of more accurate maps of magnetic anomalies. Moreover, the low-orbit satellites repeatedly measure the field at the same location many times during their lengthy missions.

Slight differences in the repeated measurements allow better control of changes in the Gauss coefficients with time, which makes it possible to observe directly the secular variation of each coefficient in the IGRF.

In order to describe the lithospheric field in even more detail, the satellite data have to be extended by data from terrestrial surveying, carried out on land, at sea, and from the air.

The terrestrial and satellite sets of data are complementary. Magnetic fields decrease in strength rapidly with distance from the source. Consequently, terrestrial surveys record preferentially the *crustal* anomalies. An airborne or marine survey made at or only a few kilometers above the Earth's surface is better able to measure shallow crustal anomalies than those caused by deep-seated sources. By contrast, satellite measurements from a low-Earth orbit are better able to define broad anomalies caused by deeper sources, such as in the upper mantle. The lithosphere is defined as consisting of the crust and the uppermost part of the mantle, which together form the rigid outer shell of the Earth. Accordingly, the terms in the spatial power spectrum of degree higher than 16 are referred to as the *lithospheric field*.

3.8   The Lithospheric Magnetic Field

The part of the spatial magnetic spectrum with harmonic degrees n larger than 16 is produced by variations in the magnetization of the rocks and minerals of the lithosphere. Many geological factors—such as volcanism or the development of tectonic structures—may cause local or regional variations in crustal magnetization. The magnetized structures, in turn, produce anomalous magnetic fields that are often much smaller in spatial extent than can be resolved from satellite altitude. A combined analysis of the CHAMP and Swarm missions resolves harmonic degrees in the geomagnetic field up to around $n = 150$. According to the relationship between harmonic degree and the horizontal extent, or wavelength, of a feature (Eq. 3.6), the lithospheric field that can be determined by satellite missions covers wavelengths shorter than 2,500 km ($n = 16$) and broader than about 260 km ($n = 150$). However, magnetic anomalies with smaller horizontal dimensions cannot be resolved from satellite measurements alone. In order to measure and interpret local and regional anomalies detailed magnetic surveys need to be carried out on land, at sea, and from the air (Chapter 2.5).

The topographic services of many countries have carried out national aeromagnetic surveys, providing coverage of continental areas. Research vessels from academic research institutions have made marine magnetic surveys of large areas of the oceans. Terrestrial surveys have often been made by private commercial companies in the search for essential and exploitable minerals, and their results are largely proprietary. Data from ground-based, airborne, and marine magnetic surveys acquired during 50 years by more than 100 institutions have been collated and compiled in the form of global maps of the world's magnetic anomalies. The lack of uniformity in the available data makes the task of modeling the lithospheric anomalies similar to that of sewing together a patchwork quilt or solving a jigsaw puzzle.

The individual surveys that must be incorporated in a model of the global crustal anomalies were conducted according to the individual goals and resources of the investigating team, and not to a coordinated plan. Consequently, formation of a global map requires integrating a patchwork of many different databases from satellite, airborne, marine, and ground-based surveys, each of which was made at a different altitude. For example, national aeromagnetic surveys were flown at different altitudes, depending on local topography; the corrections made during data processing were not consistent; some of the datasets were incomplete; and although the tracks of marine surveys were all at sea level, the sources of the marine anomalies were at different ocean depths, several kilometers below the ship's track. In order to make the data from numerous surveys compatible, many corrections must be made. Surveys made in different eras included different corrections for temporal changes in the inducing field; airborne measurements must be downward (or upward) continued to a common altitude; and marine data must be upward continued to the same altitude. The different "patches" must then be smoothly merged. Some regions have been measured with a higher density of measurements than others. This is compensated by laying a fine-meshed grid of points over the data and calculating a representative value for each point.

A recent example of a global lithospheric field model is the World Digital Magnetic Anomaly Map (WDMAM). The map was computed by reprocessing more than 25 million measurements of the field intensity obtained in more than 3,000 surveys. The computational grid was made up of 1,620,000 data values, each representative of a small block 0.2×0.2 degrees in area. The WDMAM map, version 2.0 (Fig. 3.10), shows the global distribution of lithospheric magnetic anomalies, computed for an altitude of 4 km. In order to be usable for different locations, the data are fitted with spherical harmonic functions so that they extend the model parameters from satellite missions. This resulted

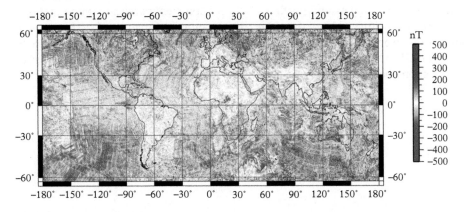

Fig. 3.10 *The World Digital Magnetic Anomaly Map, version 2, 2015. (J. Dyment, V. Lesur, M. Hamoudi, Y. Choi, E. Thébault, M. Catalan, the WDMAM Task Force, the WDMAM Evaluators, and the WDMAM Data Providers.* World Digital Magnetic Anomaly Map version 2.0—*available at http://www.wdmam.org)*

in a model of field intensity with Gauss coefficients up to degree and order 800 that uses 641,600 coefficients to represent the anomalies of the lithospheric field. The WDMAM has a nominal resolution of 2 arc-minutes, equivalent to a horizontal distance of 3.7 km, but the actual resolution varies with location depending on the available data in that region.

Magnetic anomalies over the continents and oceans have visually different appearances. Continental rocks have a very wide spectrum of ages. The youngest are freshly erupted lavas, like those formed by active volcanoes in Alaska and the western United States. The oldest rocks, with age dates of more than 4,000 Ma, are found in continental "shield" areas in Canada, Australia and South Africa. The grain sizes of the ferrimagnetic minerals in continental rocks are also very varied, ranging from coarse to very fine, which affects how easily they can be magnetized by the geomagnetic field and how well they retain a magnetization.

By contrast, the entire oceanic crust has been formed during the last 200 Myr by the vulcanism at active oceanic ridges, where global tectonic plates accrue (Chapter 5.9). Freshly erupted lava at a ridge axis has a very fine grain size and acquires a strong, stable magnetization parallel to the magnetic field in which it cools. A long stripe of strongly magnetized crust is formed parallel to the axis of the ridge, with a polarity that corresponds to that of the field. Seafloor spreading at the ridge moves the stripes apart and creates a new oceanic basement between them. If the polarity of the magnetic field changes during this process, a new stripe forms, with a remanent magnetization in the opposite direction. A strong magnetization contrast results at the margin between adjacent stripes. It produces oceanic magnetic anomalies with amplitudes of up to 1,000 nT or more, which extend along the length of the ridge.

Perpendicular to a ridge axis, the magnetic anomalies have wavelengths ranging from a few kilometers to several hundred kilometers, depending on the rate of spreading at the ridge and the rate of polarity reversals. The anomalies form a striped pattern aligned with the ridge axis that may extend for thousands of kilometers parallel to it and that is often symmetrical on either side of the ridge. The magnetic signature of oceanic magnetic anomalies is strong enough that parts of it can be measured from satellites in low Earth orbit. The CHAMP and Swarm missions are able to resolve the trends of the broader lineated anomalies.

4

The Geomagnetic Dynamo

Introduction

The interdependence of electrical currents and magnetic fields was discovered in the first half of the 19th century. The experiments of André-Marie Ampère and Hans Christian Ørsted in the 1820s demonstrated that a steady electrical current produces a magnetic field around the conductor through which it flows. Conversely, a *steady* magnetic field does *not* produce an electrical current in a conductor at rest. In 1845 Michael Faraday demonstrated that a *changing* magnetic field induces an electrical current in a conductor. This established the important phenomenon of magnetic induction. In 1873 James Clerk Maxwell unified the laws of electricity and magnetism in a set of equations that founded the classical understanding of electromagnetism. The equations explain the common nature of electricity, magnetism, and radiation, including light.

Magnetic (or electromagnetic) induction is the underlying mechanism by which the geomagnetic field is generated. In 1919 Joseph Larmor proposed that the geomagnetic field and the magnetic fields in the Sun and stars are generated by an induction mechanism that is similar to an electrical dynamo. Subsequent investigations of the origin of the geomagnetic field developed this concept further. Geochemical and geophysical information from laboratory measurements and theoretical studies provide important values for the physical properties of the core. Based on this knowledge, modern computer simulations of the dynamo mechanism are able to explain many features of the geomagnetic field.

4.1 The Concept of a Self-sustaining Geodynamo

In the early part of the 20th century, seismologists worked out the internal structure of the Earth. Measurements of the geomagnetic field at the surface and from space established that it originates in the liquid iron core. It is created by the same kind of dynamo process that converts mechanical energy into electrical energy in an electrical power station or in the lighting system of a bicycle. In these familiar examples, an electrical current is induced in a coil of wire (the conductor) by rotating it in a magnetic field. In

The Earth's Magnetic Field. William Lowrie, Oxford University Press. © William Lowrie (2023).
DOI: 10.1093/oso/9780192862679.003.0004

the Earth's outer core, the molten iron acts as the electrical conductor for a *geodynamo*. An electrical current is induced in the liquid iron as it flows through the magnetic field lines in the core. This current produces a secondary magnetic field that reinforces and sustains the original field.

A profound understanding of the currents and fields that form the geodynamo has evolved as a result of advances in data acquisition and theoretical modeling. The acquisition of scientific data underlies the theoretical models. High-quality measurements of the magnetic field are provided by magnetic observatories and satellite missions. Downward continuation from the surface where the data are acquired (the Earth's surface or a satellite's altitude) to the surface of the core, brings the data closer to their source and enables more detailed analysis. Independently, laboratory investigations of the physical and chemical properties of the liquid iron alloy—such as its composition, density, viscosity, and thermal and electrical conductivities—try to approach the extreme conditions in the core and to predict how the pressure and temperature in the core affect its physical parameters and magnetic behavior. The physical data are incorporated into the construction of theoretical models of the geodynamo that simulate the geometry and time dependence of the present field on the core surface. The simulations must ideally explain both the short-term secular variations of the geomagnetic elements and the long-term changes such as reversals of magnetic polarity.

The simulation of a self-generating geodynamo that replicates geomagnetic behavior is an exceptionally difficult challenge for mathematical geophysicists. The simulations require massive computing power, and even with a supercomputer they involve lengthy calculations. The liquid core is inaccessible, and the estimation in laboratory experiments of parameters relevant to the core may involve extrapolations with large uncertainties. For example, recent theoretical models suggested that electrical and thermal conductivities may be several times higher than previously thought. The high values have important implications for simulations of the geodynamo. If real, they would change scientists' understanding of the way in which heat leaves the core and the amount of energy available to drive the geodynamo. However, subsequent laboratory measurements have found lower values of both conductivities and support the current concept of a geodynamo powered by both thermal and compositional convection.

The real geomagnetic field is created in a space and on a time scale that can only be approximated in computer simulations. Some model values are very different in size from the real values in the core, but they allow the observed field to be simulated remarkably well. The behavior of the field over long time intervals, including the occurrence of polarity reversals, has been simulated by solving the system of geodynamo equations repeatedly in a succession of time steps.

For a self-generating dynamo mechanism in the outer core to power the geomagnetic field, several conditions must be fulfilled. The electrical conductivity of the core fluid must be large enough to support the strong electrical currents that cause the fields; the thermal conditions in the core must allow the fluid to be driven by thermal and compositional convection; and the rotation of the Earth needs to be fast enough to produce a Coriolis force that dominates the fluid motions.

4.2 Heat Transport in the Core

Heat can be transported by three mechanisms: radiation, conduction, and convection. Radiation is important in stars like the Sun, but the temperature in the Earth's core is too low for radiation to be an important mechanism in transporting thermal energy. The small amount of energy radiated from the core is absorbed before it can leave the core, thereby increasing its temperature. Heat transport by radiation can therefore be accommodated as a modification of the core's ability to transport heat by conduction.

The rate at which thermal conduction can transport heat in a material depends on a physical property of the material (its thermal conductivity) and on how the temperature changes with position (the temperature gradient). Conduction is an important mechanism for transporting heat through the solid outer shells of the crust and lithosphere as well as in the solid inner core. Thermal conduction takes place by the transfer of kinetic energy in collisions between the atoms and molecules of a material. This works well in solids, where neighboring atoms are bound to each other in a lattice and can transmit energy to each other through their vibrations. At the atomic level the laws of quantum mechanics come into force, and the vibrational energy of a crystal lattice is quantized in units called *phonons*. Inside the Earth, conduction by phonons is an effective mechanism for transporting heat through the solid mantle and crust. Thermal conduction can also take place by the motion of free electrons (i.e., those not bound to a particular atom). They can drift through a material, moving down the temperature gradient and transporting thermal energy. This form of conduction can take place in liquids as well as solids, so it may contribute to heat transfer in the core.

Thermal conduction is a viable method of transporting heat in a liquid and contributes significantly to removing heat from the core. However, it does so less efficiently than convection, in which freely moving molecules transfer heat bodily from a hot region in the liquid to a cooler one. As a result, convection is the most important mechanism of heat transfer in the liquid outer core. It takes place by two mechanisms. The better-known form is thermally driven and may be observed in any kitchen when soup is heated in a pot. Thermal convection takes place in a fluid when the temperature gradient exceeds a critical value, known as the *adiabatic gradient*. If it is being heated from below, the hot fluid expands and buoyancy forces cause it to rise. Colder fluid sinks to replace it, renewing the cycle. The temperature in the outer core is highest at the boundary to the solid inner core. At this interface, the solidification of liquid iron releases a latent heat of melting, which augments the existing heat in the outer core. Thermal convection transports the heat outward from the surface of the inner core to the core–mantle boundary.

The core supports an additional form of convection, which results from changes in the chemical composition of the liquid iron alloy. It is a consequence of the solidification of liquid iron at the inner core boundary. With increasing depth in the outer core, the temperature continues to rise, but the pressure increases even more rapidly. Eventually, the pressure becomes so high that the iron cannot remain liquid. At a depth of 5,150 km, although the temperature is higher than 6,200 K, the liquid iron solidifies at the surface of the inner core. Seismology indicates that the inner core is about 5% more dense than the outer core. The density difference is thought to be due to small amounts of lighter elements—such as sulfur, silica, and oxygen—which are left behind in the outer core during solidification. The light elements are less dense than the liquid iron, and so they are buoyant and rise to the surface of the outer core, carrying heat with them. This sets up a convection cycle that is driven by the slight difference in chemical composition across the boundary between the inner and outer core. Computer models show that this compositional convection is probably more important than thermal convection in powering the processes that produce the geomagnetic field.

Heat escaping from the liquid core into the solid mantle crosses the physical core–mantle boundary layer slowly, so that, as well as marking a change in composition, the CMB is also a thermal boundary layer. It causes a buildup of heat, so that a steep temperature gradient develops in the outermost core. In the deep mantle, the high temperature mobilizes defects in the mantle minerals, which migrate through the crystalline structures and result in a transport of mass and heat. This results in thermal convection in the viscoelastic mantle. It is a slow process, taking on the order of 200 Myr for a complete turnover of the mantle.

4.3   The Coriolis Force Due to the Earth's Rotation

The Earth's rotation produces a centrifugal force that modifies the planet's shape, causing it to be spheroidal (i.e., flattened at the poles), rather than perfectly spherical. As a result, the acceleration of gravity is about 0.5% stronger at the poles than at the equator. Viewed from above the north pole, the rotation is in an anticlockwise sense, so that, to an external observer, the surface moves from west to east.

The rotation also causes additional forces on any mass that is moving across the Earth's surface. In general, the velocity of the mass has both a northward and an eastward component. The eastward component adds to the linear velocity of the rotation and causes the centrifugal force to increase. The direction of the increase is perpendicular to the rotation axis, and its magnitude varies with latitude; it is maximum at the equator and zero at the poles. It can be resolved into a vertical component that acts radially and a horizontal component that acts parallel to the Earth's surface. The vertical component is called the *Eötvös* force. It modifies any measurement of gravity that is made on a moving vehicle, such as a research vessel or aircraft. The horizontal component is called the *Coriolis* force. It acts at right angles to the original velocity in such a way that it deflects a moving mass toward the right of the direction of motion in the northern hemisphere (Fig. 4.1) and to the left in the southern hemisphere. The Coriolis force constrains wind patterns in the atmosphere to form cyclones and anticyclones, the "circular" patterns of flow about centers of low and high pressure, respectively, that are familiar from weather forecasts. In the liquid core, it has a strong influence on the generation of the Earth's magnetic field.

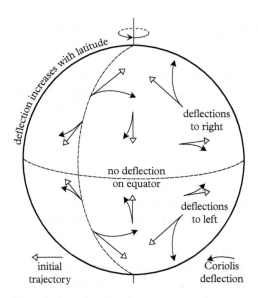

Fig. 4.1 *The path of a horizontally moving object at the Earth's surface is deflected by the Coriolis acceleration due to the Earth's rotation. The deflection is to the right in the northern hemisphere and to the left in the southern hemisphere.*

4.4 Magnetohydrodynamics and the Frozen-Flux Concept

The ability of a material to conduct an electrical current is measured by its *electrical conductivity*, σ. This physical property extends over an extremely wide range of values. At the low end, it may be so small that a material does not conduct electricity at all, and it is called an insulator. Wood, rubber, and plastics are familiar examples. At the high end there are special materials that offer so little resistance to an electrical current that it can flow indefinitely; this type of material is known as a *superconductor*. Between these extremes metals such as copper, silver, and aluminium are classified as good electrical conductors, with conductivities in the range 3.8–6.3×10^7 S/m. For comparison, the electrical conductivity of iron at room temperature is around 10^7 S/m.

Estimates of the electrical conductivity in the Earth's liquid outer core depend on the temperature assumed at the inner core boundary. Different compositions of the molten iron alloy in the outer core give consistent estimates in the range 1.35–1.55×10^6 S/m. The electrical conductivity of the inner core is thought to be 18–25% higher, at about 1.8–2.0×10^6 S/m. The mantle has a much lower electrical conductivity: some models suggest that it is similar to that of crustal rocks, on the order of 0.01 S/m, so that it is often treated as an insulator. The liquid nature of the outer core and its high electrical conductivity are key aspects of how the magnetic field is generated.

In 1942 a Swedish physicist, Hannes Alfvén, formulated an important concept, which has come to be known as the *frozen-flux* theorem. It explains how a magnetic field becomes tied to the flow of an electrically conducting fluid that passes through it, when the electrical conductivity is high enough. The theorem helps to explain how the geomagnetic field is generated in the Earth's liquid core. Frozen-flux behavior depends on two properties of the conducting fluid: its electrical conductivity and the velocity of its motion relative to the magnetic field.

Alfvén's explanation of the interaction between a magnetic field and a moving, electrically conducting fluid is fundamental to *magnetohydrodynamics*. This field of science combines Maxwell's laws of electromagnetism with the laws of fluid dynamics. The latter are described by a set of equations known as the *Navier-Stokes* equations, which express mathematically the flow of a viscous fluid. Magnetohydrodynamics is an advanced topic in theoretical physics and is beyond the scope of this book. However, we can learn how it applies to the geomagnetic field by taking a quick look at the basic electromagnetic equation for a moving fluid.

When a conductor moves with velocity $\mathbf{v}$ through a magnetic field $\mathbf{B}$, an electrical current is induced in the conductor. In this situation, the law of magnetic induction becomes modified to include a term describing the effect of the motion. The equation for the changing magnetic field is

$$\frac{\partial \mathbf{B}}{\partial t} = \frac{1}{\mu\sigma}\nabla^2\mathbf{B} + (\nabla \times \mathbf{v} \times \mathbf{B}) \tag{4.1}$$

In this equation the variables t, $\mathbf{v}$, and $\mathbf{B}$ represent time, the velocity of the flow, and the magnetic field, respectively. The properties of the liquid iron are its electrical conductivity, σ, and its magnetic permeability, μ, which in the core is close to the permeability of free space, $\mu_0 = 4\pi \times 10^{-7}\,\mathrm{N/A^2}$. The other symbols in the equation describe mathematical operations. The term on the left of Eq. (4.1) represents the changing magnetic field. The terms on the right represent two different mechanisms that cause the magnetic field to change. Each mechanism plays an important role in the dynamo process by which the geomagnetic field is generated and sustained.

Consider first the situation if the liquid is stationary, that is, $\mathbf{v} = 0$. The term in brackets on the right of Eq. (4.1) becomes zero. The first term on the right depends inversely on the electrical conductivity, σ, and represents the resistance of the conductor to an electrical current. The molten iron in the Earth's core is a good electrical conductor ($\sigma \sim 10^6$ S/m) compared to the overlying silicate mantle ($\sigma \sim 0.1$ S/m), but its resistance is not negligible. The current loses energy in overcoming the resistance, which causes the magnetic field to decay exponentially with time. The process functions in the same way that heat diffuses through a material when it cools, and for this reason the term is called the *diffusion* term. In Eq. (4.1) it causes the magnetic field in the outer core to decay with time. The time τ for the field to drop to 1/e of its initial value (where e = 2.718

is Euler's number, the base of natural logarithms) is called the *diffusion time* and is given by

$$\tau = \mu_0 \sigma R^2 / \pi^2 \qquad (4.2)$$

On inserting appropriate values for the parameters, the diffusion time of the magnetic field in the core is found to be on the order of 20,000 years. However, paleomagnetic evidence indicates that the Earth has had a magnetic field for hundreds of millions of years, so in order to generate the magnetic field a dynamic source is needed.

The second term on the right of Eq. (4.1) depends on the flow velocity, **v**. It is a result of *motional induction*, whereby an electrical current is induced in a conductor that is moving through a magnetic field (Chapter 1.4, Eq. 1.3). In turn, the induced current produces a secondary magnetic field. If the flow is fast enough, or if the electrical conductivity is high enough, the rate at which the magnetic field is induced by the movement of the conductor can be more important than the rate at which the field can decay by diffusion. Under the right conditions, the magnetic field generated by the motion of the core fluid reinforces and maintains the original magnetic field. This is the mechanism of the self-sustaining *geodynamo* that produces the geomagnetic field in the fluid outer core.

4.5 The Dynamo Model for the Origin of the Internal Magnetic Field

Buoyancy forces drive the convective flow of liquid iron in the Earth's core, whereas the viscosity of the liquid and the loss of heat by diffusion resist the flow. The relative importance of the forces driving and opposing convection is expressed by a dimensionless quantity called the Rayleigh number, which takes into account the physical properties of the core: gravity, density, coefficient of thermal expansion, temperature gradient, thermal conductivity, viscosity, and the thickness of the convecting volume. Estimates of the Rayleigh number in the Earth's outer core are larger than 10^{27}. This is greatly in excess of the critical value needed for convection to begin, which is on the order of 10^3. Thus, the conditions in the outer core strongly favor heat transfer by turbulent convection.

In the magnetohydrodynamic equation (Eq. 4.1), the motional induction term expresses how the magnetic field changes as a result of the fluid flow, while the diffusion term describes how the field decays. The relative importance of the two terms is expressed by their ratio, which defines another dimensionless quantity called the magnetic Reynolds number. This number is found to be of the order of 1,000 in the outer core. This means that to a first approximation, and when considering short time intervals, the diffusion term can be neglected. The frozen-flux theorem is then applicable, and the magnetic field lines are carried along, following the geometry of the flow.

An interplay between three forces controls the flow of the liquid iron in the core. Buoyancy forces drive the flow, transporting heat radially outward from the inner core boundary, while the Earth's rotation produces strong Coriolis forces perpendicular to the flow direction that twist the flow pattern. The motions of the liquid iron distort magnetic field lines and are opposed by Lorentz forces. The forces driving the flow are also opposed by the drag of the fluid's viscosity. The ratio of the viscous drag to the Coriolis force defines another dimensionless parameter, the Ekman number. For large-scale flow on the size of the outer core, the Ekman number turns out to be extremely small, on the order of 10^{-15}, which implies that the large-scale circulation of the core fluid is not restricted by the viscosity.

The buoyancy and Coriolis forces mold the flow lines of the core fluid into tall, thin columns, parallel to the rotation axis. The fluid motion inside the columns has a helical (or spiral-like) configuration (Fig. 4.2). The helicity—the direction and pitch of the spiral—has the opposite sense in adjacent columns, and it changes sense in each column at the equator. The most important region of the core for this activity is outside the *tangent cylinder*. This is an imaginary cylinder, with its axis along the rotation axis and with a radius equal to that of the inner core, so that the cylinder just touches the inner core at the equator. The intersection of each end of the cylinder with the surface of the core is a circle. Computations of the magnetic field on the core surface have shown that magnetic flux entering or leaving the core is concentrated in two patches outside the circle and that the flux is low inside it.

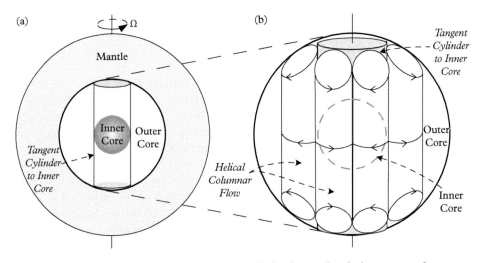

Fig. 4.2 *(a) The tangent cylinder is an imaginary cylinder that touches the inner core at the equator and is parallel to the rotation axis. It divides the patterns of flow in the outer core into two regimes. (b) Outside the tangent cylinder, the Coriolis acceleration dominates the flow, forcing it to form cylindrical columns in which the flow has a helical configuration.*

The helical motion of the conducting fluid twists the magnetic field lines, which break and reconnect, giving rise to toroidal and poloidal magnetic fields in the core. Toroidal field lines form around the rotation axis and have no radial component; they are parallel to the surface of the core and are thus confined to the core. The strength and configuration of a toroidal field cannot be measured outside the core and must be estimated from models. The interactions of toroidal field lines with the helical flow within the columns cause them to twist and reconnect. This action produces poloidal magnetic fields that escape from the core and combine to produce the field measured at the surface of the Earth. The Coriolis force that shapes the helical flow within the columns is a product of the Earth's rotation, and therefore causes the geometry of poloidal fields outside the Earth to favor the rotation axis.

The parameters used to simulate the field are the physical properties of the core fluid, the rotation of the planet, the temperature difference between the inner core boundary and the core–mantle boundary, the flux of light elements released at the inner core boundary, and the dimensions of the core. The regime in the outer core is much more complex than in the description above, which, however, gives an impression of the interacting processes involved in generating the field. A simulation, which may take hours or days of expensive time on a supercomputer, is not a real-time reproduction of the actual geodynamo, which functions on a much longer time scale. In order to make a functional simulation, there is a disparity between the control numbers used in the models and their real values in the core. For example, to carry out a complex computer simulation in a reasonable length of time with a supercomputer, an Ekman number $\sim 10^{-6}$ is used rather than the real value of $\sim 10^{-15}$ in the core. These problems are less significant than the spectacular success of the simulations, which have contributed greatly to a theoretical understanding of the core's dynamics.

The first dynamically consistent simulation of a self-sustaining dynamo was a three-dimensional model (Glatzmaier & Roberts, 1995). It shows smooth poloidal field lines outside the core, which change at the core–mantle boundary to a complicated system of toroidal field lines inside the core (Fig. 4.3). The simulation was carried out for a long enough time that the model spontaneously underwent a reversal of polarity. Many such simulations of the present field and of polarity reversals have been carried out since this pioneering study. Although some problems remain to be resolved, the manner by which the geomagnetic field is generated is now basically understood.

4.6 The Magnetic Influence of the Inner Core

Seismology has established that the inner core behaves as an elastic solid for the passage of seismic waves and that it has a radius of 1,221 km. The composition and state of the inner core are not yet fully understood, and so they are targets for ongoing research. It is believed to consist of solid iron alloyed with about 2% of light elements, which is a smaller fraction than in the outer core. Because many lighter elements remain in the outer core when the inner core solidifies, the density increases at the inner core boundary from $\sim 12,100$ kg/m^3 to $\sim 12,800$ kg/m^3. For comparison, the density of solid iron at the Earth's surface is $\sim 7,800$ kg/m^3; the density of water is 1,000 kg/m^3.

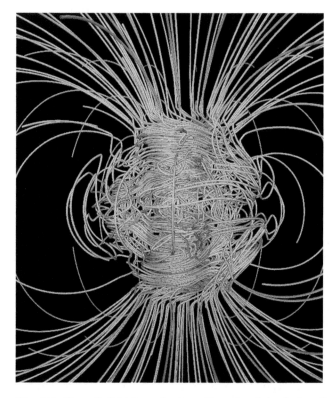

Fig. 4.3 *Magnetic field lines of a three-dimensional simulation of the geomagnetic dynamo. (G. A. Glatzmaier and T. Clune, Compu-tational aspects of geodynamo simulations,* Computing in Science & Engineering, *2, 61–67, 2000)*
Reprinted with permission from the Institute of Electrical and Electronics Engineers (IEEE).

The solid condition of the inner core means that it can only flow by the migration of defects, as is the case for flow in the solid mantle. This viscoelastic type of flow is very slow, and hence flow in the inner core does not participate directly in the geodynamo process. However, the poloidal geomagnetic field is present in the inner core, where, because of the virtual absence of flow, the magnetic field can only change by diffusion. Estimated values of the thermal and electrical conductivity at the top of the inner core are ~50% and ~20% higher than in the fluid outer core. A change of the field that penetrates the inner core therefore has an even longer diffusion time (Eq. 4.2) than in the outer core. Consequently, the inner core is thought to have a stabilizing influence on the core field by dampening rapid changes.

Computer simulations have indicated that the inner core may have an important func-tion in the occurrence of geomagnetic polarity reversals. Simulations have shown that in the absence of an inner core, the field attempts to change polarity frequently and chaotically. The paleomagnetic record of polarity is interspersed with many large angular departures of the magnetic pole from a polar region. Instead of completing a polarity reversal, the pole usually returns in a geologically short time to the original polar region. The temporary displacement is called a magnetic *excursion*. It is possible that the inner core is the reason for the failure of the field to reverse fully. In a complete reversal, both the inner core and the outer core must change polarity. Most attempts of the outer core field to reverse polarity may be unsuccessful because the field in the inner core does not have enough time to adjust to the new configuration before the outer core again reverses and returns to its original polarity, thereby canceling the initial attempt and leaving merely an excursion as evidence of the activity. If this is the case, the inner core provides stability to the reversal process and supports the observed structure of polarity reversal sequences, in which long intervals of constant polarity are interspersed at irregular intervals by successful reversals.

The inner core evidently plays an important role in the self-sustaining geodynamo process. It grows in size at the expense of the outer core, and, as it solidifies, it leaves behind light elements, and so it powers the compositional convection in the outer core. Moreover, the diameter of the inner core determines the diameter of the tangent cylinder, outside of which the helical columnar flow takes place, which is an important factor in generating the main field (Fig. 4.2). Determining when the inner core started to develop is therefore of interest for understanding core dynamics in the early Earth.

4.7 The Magnetic Field at the Core–Mantle Boundary

The geomagnetic field can be examined in greater detail closer to its source. This is achieved by downward continuation of the field from the Earth's surface to the core–mantle boundary (Chapter 3.3). The process amplifies details in the higher degree part of the field. A map of the field at the CMB provides information about the pattern of magnetic flux leaving and entering the core (Fig. 4.4). In contrast to the field at the Earth's surface, the field at the CMB is less regular and is characterized by large regional anomalies. The general picture is one of magnetic flux that leaves the core surface in the southern hemisphere and returns in the northern. The flux does not cross the CMB uniformly but rather is concentrated in distinct patches.

The magnetic flux in the northern hemisphere is dominated by two stationary positive flux patches—where flux enters the core—located in high latitudes under northern Canada and Siberia. In the southern hemisphere, patches of negative flux—where flux leaves the core—lie over the eastern and western margins of Antarctica. The concentrations of flux are thought to be related to the locations of the helical convection columns in the core outside the tangent cylinder.

The field at the core surface includes several flux patches in which the flux has the opposite direction to the background dipolar field. Surprisingly, a patch of reverse flux is located in the vicinity of the north geographical pole. It is a persistent feature in analyses

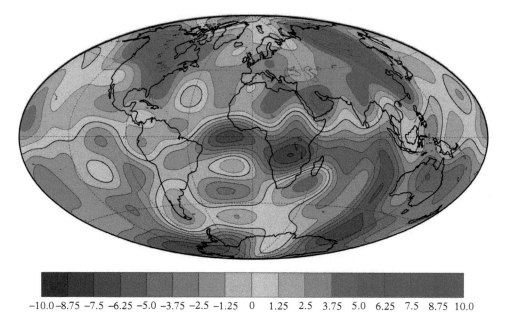

−10.0 −8.75 −7.5 −6.25 −5.0 −3.75 −2.5 −1.25 0 1.25 2.5 3.75 5.0 6.25 7.5 8.75 10.0

Fig. 4.4 *The vertical component of the geomagnetic field at the surface of the core in 1990. Blue color indicates downward direction of the field (i.e., field lines entering the core), and orange color indicates upward direction (i.e., field lines leaving the core). The intensity scale is in units of 10^5 nT. (After Fig. 31 in A. Jackson and C. Finlay, Geomagnetic Secular Variation and its Applications to the Core,* Treatise on Geophysics, *2nd edition, chief editor G. Schubert, Elsevier, Amsterdam, 2015. Vol. 5,* Geomagnetism, *editor M. Kono, pp. 137–184. With permission from Elsevier)*

of the historic field of the past four centuries. In the southern hemisphere, a broad patch of reverse flux extends across the Southern Ocean from South America to South Africa. It is responsible for the weakening of the field intensity at the Earth's surface that has become known as the South Atlantic Anomaly (Chapter 2.8). The reverse flux patches are ascribed to distortions of the field caused by local irregularities in the flow of the liquid core.

In addition to being evident in observations of the modern field made by satellites and at ground observatories, the flux patches at the core surface have also been interpreted from historic measurements of the field made by mariners. The compass readings of magnetic declination along the tracks of their journeys over the past four centuries provide a large archive of field directions. Augmented by data from magnetic observatories for the time since 1840, the navigators' records have been analyzed by spherical harmonic modeling and have been portrayed as global maps of the magnetic field on the core surface at different ages. By comparing the global maps for different ages, the historic secular variation in the field can be followed in detail at the core surface. The westward drift of the nondipole field is seen to be related mainly to the motion of flux patches located in a band of latitudes within 20° on either side of the equator. Slight variations in the strengths and positions of the strong flux patches in high latitudes may cause eastward as well as westward drift. A computer model of the field for the past 3,000 years based on archeomagnetic data indicates that at times the field drifted westward and at other times eastward (Dumberry & Finlay, 2007). Westward drift has characterized the drift from about AD 1400 until the present, but the drift was mainly eastward between AD 800 and 1400.

4.8 Archeomagnetic Secular Variation of Paleointensity

Changes in the intensity and direction of the magnetic field in the geological past are known as *paleomagnetic* secular variations (PSVs) or, more briefly, paleosecular variations. They are calculated by analyzing the remanent magnetizations of radiometrically dated rocks and minerals following the practices of paleomagnetism (Chapter 6). The same methodology makes it possible to examine the magnetic field recorded in the remanent magnetizations of archeological artefacts (e.g., fired clay pots), as well as in historic lava flows and unconsolidated sediments deposited in lakes and the oceans. The results are called *archeomagnetic* secular variations. The history of human existence is too short for the field to undergo major differences in direction; the current changes in dipole and nondipole components are presumed to have been present throughout human history. However, archeomagnetic results have shown quite large changes in paleointensity throughout this time, although the data have a large amount of scatter.

The intensity B of the magnetic field of a dipole varies with co-latitude θ (Fig. 4.3 and Eq. 3.2). In order to compare a measurement of paleointensity at a certain location with results from other locations, it is therefore necessary to take the different co-latitudes θ of the sampling sites into account. The latitude or co-latitude at which a rock was magnetized can be calculated from the inclination of its magnetization by assuming the geocentral axial dipole hypothesis, which is the central tenet of paleomagnetism (Chapter 6.2). The measured paleointensity is then converted to the strength—called the virtual dipole moment (VDM)—of the geocentric axial dipole that would produce the measured field B at magnetic co-latitude θ. The relationship between VDM and B for a geocentric dipole field is obtained by rewriting Eq. (3.2) in the form

$$VDM = \left(\frac{4\pi R^3}{\mu_0} \right) \frac{B}{\sqrt{1 + 3\cos^2\theta}} \qquad (4.3)$$

where R is the Earth's radius and μ_0 is the magnetic constant, as before.

A laboratory measurement of paleointensity involves multiple cycles of heating and cooling a sample. The procedure is designed to recognize alteration of the sample, or possible remagnetization in spurious fields, which can occur in the thermal cycling. The quality of paleointensity results in the archeomagnetic database is assessed by 10 quality criteria. Unfortunately, paleointensity data are of variable quality. In an evaluation of more than 3,500 results representing the strength of the dipole moment in the past

7,000 years, only 110 results were found to satisfy all 10 quality criteria. This is partly because, in investigations that were made decades ago, less stringent laboratory methods were employed than are customary today. By accepting older studies in which alternative reliability tests were applied, the number of acceptable data increases to about 50% of the database. Nevertheless, the geographic distribution of reliable sites in the depleted database is uneven: almost 75% come from western Eurasia, and there are few results from the southern hemisphere. The sites are also poorly distributed in time, with a predominance of data in the most recent 1,000 years (Fig. 4.5, lower panel).

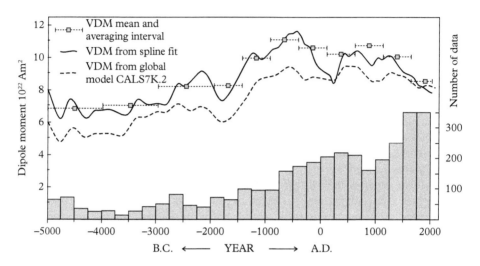

Fig. 4.5 Upper: *Variation of the magnetic moment of the geomagnetic dipole during the last 7000 years, based on selected data from the ArcheoInt paleointensity database.* Lower: *The age-distribution of the data, grouped in 250-year bins.* (Data sources: *M. Korte and C. G. Constable, Centennial to millennial geomagnetic secular variation.* Geophysical Journal International, *167, 43–52, 2006; A. Genevey, Y. Gallet, C. G. Constable, M. Korte, and G. Hulot, ArcheoInt, An upgraded compilation of geomagnetic field intensity data for the past ten millennia and its application to the recovery of the past dipole moment.* Geochemistry, Geophysics, Geosystems, *9, 43–52, 2008)*

The depleted archeomagnetic database has been used to investigate the changes in strength of the dipole field during the past 7,000 years. The temporal variations in paleointensity were analyzed in three ways (Fig. 4.5, upper panel). In the simplest analysis, mean values of VDM were calculated for fixed time intervals of 1,000 years between 5000 BC and 2000 BC, and for time intervals of 500 years after 2000 BC. In a second analysis, a moving average was calculated by fitting spline functions to all the data (solid curve). In a third analysis (dashed curve), a continuous global field model was calculated by fitting spherical harmonics to the geographic distribution of values; the time dependence was represented by fitting a smooth mathematical function to the successive values of each Gauss coefficient. The strength of the VDM was then computed from the dipole coefficients of degree $n = 1$ using Eq. (3.6).

The estimated VDM of the field exhibits large variations in each of the three types of analysis. The first two analyses agree well with each other. The VDMs derived from the global model are systematically about 20% lower because this method uses only the extracted dipole coefficients and is free of nondipole components. Prior to about 1500 BC, the archeomagnetic dipole moment was a few percentage points weaker than that of the present field (around 7.7×10^{22} A/m²). From ~1500 BC until AD ~500 the intensity was about 25–30% stronger than the present field.

The paleointensity record during the past 7,000 years exhibits large fluctuations on a millennial time scale. These changes indicate that the recent decrease in intensity of the dipole field is not unusual; therefore, it does not imply that a polarity reversal is necessarily imminent. Although decreased paleointensity does accompany a polarity reversal, the paleomagnetic record contains numerous intensity fluctuations that are unrelated to reversals or excursions.

4.9   The Geomagnetic Field in the Early Earth

The modern geodynamo is powered by thermal and compositional convection. It is uncertain at what stage of the Earth's evolution these processes began. Some modeling experiments have suggested that the thermal conductivities of liquid iron alloys at the temperature and pressure of the outer core are much higher than previously accepted values. A high thermal conductivity allows more heat to escape by conduction. If enough heat is transported by conduction, less is available to drive thermal convection, and there might not be enough to power a geodynamo in the liquid core. The onset of compositional convection is therefore an important factor in determining when the modern geodynamo originated.

Radioactive sources of heat are not thought to be important in the core. The difference between heat entering the liquid core at its boundary with the inner core and the outflow of heat across the core–mantle boundary provides the energy that powers the geodynamo. The pattern of heat flow at the core–mantle boundary therefore plays an important role in the energy balance of the liquid outer core. The thickness of the D″ layer above the CMB is spatially variable, and changes could influence the transport of heat out of the core. Numerical models suggest that variations in mantle convection could also influence the behavior of the geodynamo, for example, by modulating the frequency of polarity reversals.

The Earth's solid inner core also plays an important role in the geodynamo because the compositional form of convection depends on the release of lighter elements as the core fluid solidifies at its surface. Without an inner core, the concept of a tangent cylinder has no meaning; the different forms of convection inside and outside the cylinder that are factors in generating the present field would not exist. If the early Earth lost much of its heat by conduction, thermal convection might have been inadequate to power the complex mechanisms involved in generating a dipolar magnetic field.

Very old Precambrian rocks have remanent magnetizations that are carried by tiny grains of magnetite, as is the case in younger rocks. The existence of these ancient magnetizations shows that some kind of magnetic field existed even in the earliest Archean eon, although the data are too fragmentary to assess whether the field was dipolar or multipolar. If an inner core had not yet nucleated, a different mechanism than the current geodynamo would be needed to explain this evidence for an early field.

Despite many investigations of the early magnetic field recorded in the magnetization of ancient rocks and minerals, paleomagnetic research has not yet been able to identify unambiguously when the inner core started to grow. Estimates range from an "old" age around 4000 Ma, early in the Archean eon, to a "young" age around 565 Ma in the Ediacaran, which is the latest period in the Precambrian. This uncertainty clouds scientific understanding of when and how the Earth acquired its modern geomagnetic field.

5

The Magnetism of the Earth's Crust

Introduction

In 1908, a Croatian seismologist, Andrija Mohorovičić, noted that seismic P-waves from an earthquake near Zagreb arrived at some recording stations earlier than expected. From this information he inferred the existence of a deeper layer in which the waves could travel faster than those that arrived directly from the earthquake. This was the earliest seismic evidence of the boundary, now called the Mohorovičić discontinuity, that separates the Earth's crust from the underlying mantle. The crust consists of a thin heterogeneous layer of different rock types. Its thickness averages less than 1% of the Earth's radius but is very variable. Oceanic crust is composed of dark igneous rocks such as basalt and gabbro, covered by a thin sedimentary layer and is commonly only 5–10 km thick. It is constantly renewed at oceanic ridges and is younger than about 200 Myr in age. Continental crust is less dense than the oceanic crust and consists of sedimentary rocks and igneous rocks that are more granitic in composition than oceanic rocks. It is thicker than oceanic crust, measuring around 20–80 km, and is much older. The oldest continental rocks are found in cratons, which formed the stable cores of ancient continents. They date from the Archean eon in the early Precambrian and are more than 4,000 Myr old.

The magnetic signatures of oceanic and continental crust have been studied in detail since the Second World War in airborne and marine magnetic surveys. The surveys locate, describe, and interpret the origin of magnetic anomalies that arise from differences in crustal magnetization. The crustal magnetic anomalies on continents are largely, but not entirely, due to magnetizations induced by the present field. They provide information that helps government scientists to understand the geology of their countries, and they are used by commercial companies in the search for mineral resources. The thin magnetized layer of oceanic crust is characterized by strong positive and negative anomalies that form long, lineated structures parallel to oceanic ridges. The magnetic lineations are due to alternately polarized remanent magnetizations acquired when the oceanic crust forms at the ridges. Their interpretation has revealed the relative motions of lithospheric plates and has contributed greatly to an understanding of geological processes that are related to global plate tectonics.

The Earth's Magnetic Field. William Lowrie, Oxford University Press. © William Lowrie (2023).
DOI: 10.1093/oso/9780192862679.003.0005

5.1   Physical Properties of the Crust and Mantle

By definition, the lithosphere comprises the crust and the uppermost part of the mantle. It forms a hard, rigid outer shell to the Earth. When subjected to a small stress, the rocks and minerals in the lithosphere deform by proportionately small changes of shape. At first, the deformation is that of an *elastic* body. This means that the shape returns to its original state after a stress is removed, so that the stress does not cause a permanent deformation. However, if subjected to stress above a critical value (called the elastic limit or yield stress), a rigid material deforms in a *brittle* manner, changing shape abruptly and permanently. Instead of stretching it ruptures.

The rocks in the crust and upper lithosphere react to stress in this way. When a rock is suddenly subjected to a high stress that exceeds the elastic limit, it fractures abruptly. This occurrence manifests as an earthquake, which produces elastic vibrations that travel through the body of the Earth in the form of seismic waves. However, stress that is applied for a long time can have a different effect. Long-lasting high stress causes defects in the crystal structure of a rock to move, so that the "hard" rock deforms slowly by flowing like a very viscous fluid. This is known as a *plastic flow*. The type of deformation that takes place by this mechanism is called *ductile* deformation.

The elastic properties of rocks alter with increasing temperature and pressure, and as a result the material in the Earth's interior becomes ductile with increasing depth, as the cracks and defects in minerals become sealed by the increasing pressure. Over short time intervals—such as during the quick passage of seismic waves—the mantle behaves like an elastic solid, but over long periods of time, it deforms by flowing like a very viscous liquid. The mechanism that enables long-term flow in the mantle is called plastic deformation; it takes place slowly over time scales of millions of years. The rheology and composition of the Earth's deep interior change abruptly at the boundary between the solid silicate mantle and the liquid iron core, at a depth of 2,891 km. The boundaries between the crust and mantle, and between the mantle and core, are characterized by dramatic changes in the speed of seismic waves. The travel times of these waves have enabled seismologists to form a detailed understanding of the structure of the deep interior of the planet.

The Earth's crust is composed of crystalline minerals, most of which are so weakly magnetic that they can be neglected. A small fraction of the minerals, in particular the oxides and sulfides of iron, are comparatively strongly magnetic and are generally responsible for the crustal magnetization, except in some localized orebodies. Though crystalline, these minerals possess magnetic properties similar to metallic iron, and their type of magnetism is loosely referred to as ferromagnetism (Chapter 5.3). The ferromagnetic (*s. l.*) minerals have strong magnetizations below a critical temperature (called the Néel temperature). The magnetic properties of ferrimagnetic minerals in the Earth's crust become weaker with increasing depth due to the accompanying increase in temperature. Above their Néel temperatures, the magnetic minerals become paramagnetic, which means that they become essentially nonmagnetic. The most important ferrimagnetic minerals have Néel temperatures in the range from 200°C to 675°C. The depth at which these temperatures are reached determines the thickness of the magnetized surface layer. It depends on the local temperature gradient, which is around 20–30 °C/km near to the surface. Depending on the type of crustal rock, the magnetic minerals it contains, and the local temperature gradient, the surface layer of the Earth that can be magnetized by the geomagnetic field has a variable thickness. In places, the magnetized layer may be thicker than the thickness of the crust alone and may comprise both the crust and the uppermost part of the mantle.

The deeper parts of the lithosphere and the rest of the mantle are too hot to sustain the ferrimagnetism found in some crustal magnetic minerals. The magnetic behavior of the mantle is classified as paramagnetic or diamagnetic. These properties are generally very weak, and the mantle does not make a significant contribution to the global magnetic field. However, the thermal and electrical properties of the deep mantle may be able to affect the generation of the geomagnetic field by modulating the outflow of heat from the core.

5.2 Crustal Rock Types

There are three main categories of crustal rocks, classified according to the way they form: igneous, sedimentary, and metamorphic. The method of formation determines the magnetic properties of the rocks and, by this means, the magnetic signature of the crust and lithosphere. Most rock-forming minerals are effectively nonmagnetic, but they are interspersed with a dilute concentration of strong ferrimagnetic minerals. In an igneous rock such as a lava, their concentration can be relatively high, sometimes exceeding 1%, but in a sedimentary rock it is commonly less than 0.01%.

Igneous rocks form when molten magma from deep in the Earth rises toward the surface. Often the magma does not reach the surface but intrudes overlying rocks to form steeply inclined dykes or flat-lying sills. Some large intrusions, called batholiths, are more than 100 km across and extend to 20 km in depth. The intruded magma cools slowly and forms coarse-grained rocks such as granite and diorite. The mineral grains in these rocks have time to increase in size, and hence intrusive igneous rocks are relatively coarse-grained. If the magma reaches the surface, it may be extruded in a volcanic eruption as a molten lava, which flows across the surface, cooling rapidly until it becomes too viscous to flow further. The lava solidifies within minutes to a few days, so the minerals in the lava have little time to crystallize and grow. The result is a fine-grained igneous rock. An important example is basalt, which forms a strongly magnetic rock. Igneous rocks can be strongly magnetic: magnetic anomalies associated with oceanic basaltic rocks can be measured from space.

Sedimentary rocks are formed when unconsolidated sediments are compressed and then cemented. Some sediments are made of minerals that form chemically by precipitation from an aqueous solution. In other sediments, the grains may consist of rock fragments and minerals that were weathered and eroded from continental rocks and subsequently transported into a depositional basin. Classified as detrital grains, they are transported by wind or water and may be deposited as a sediment on the land surface or in a collection basin, such as a lake or an ocean. The method of transport winnows the sedimentary particles by size. For this reason, some deep-sea sediments that are deposited far from land are very fine-grained, whereas sediments deposited in rivers or on ocean margins form coarser-gained sands and sandstones. After cementation and diagenesis—which is a process of chemical and physical modification of the original mineralogy—the sediment hardens into a sedimentary rock. Sedimentary rocks such as sandstone and limestone consist largely of quartz or carbonate minerals, respectively. They contain very low concentrations of ferrimagnetic minerals and are only very weakly magnetic.

Metamorphic rocks may have originated as igneous or sedimentary rocks, but they were altered subsequently by high temperatures and high pressures. These conditions are produced by deep burial or by tectonic forces that deform the rocks enough to alter their mineralogy. Metamorphism can profoundly alter the magnetic properties of a rock; it may destroy their original magnetism, or, conversely, it may cause them to become strongly magnetic. As a result, the magnetic properties of metamorphic rocks can be as strong as igneous rocks or as weak as sedimentary rocks, depending on their original composition and the history of heating and deformation they have experienced.

5.3   Types of Magnetism in Minerals

In a magnetic field, the atomic magnetic moments in a mineral experience a torque that tries to align them with the direction of the field. A statistical net alignment of magnetic spins is produced, which is proportional to the strength of the magnetic field and is known as an *induced magnetization*. The ratio of the magnetization to the inducing field is called the magnetic *susceptibility* of the mineral. It is a measure of how easily a material can be magnetized by an applied field.

All minerals react to a magnetic field, displaying a very weak response called *diamagnetism*. This can be explained in terms of the Bohr model of the atom. When the electron orbitals around an atom contain an even number of electrons, they adjust to an applied field by changing their orientations so as to oppose the field. This causes a diamagnetic material to acquire a magnetization in the opposite direction to an applied field. The magnetic susceptibility of a diamagnetic material is therefore negative. The diamagnetic effect is very weak and can only be observed if each of the electron spins in an atom is paired with another electron spin. Many common rock-forming minerals are diamagnetic. Important examples are quartz in igneous rocks and sandstones, and calcium carbonate in limestones. This weak type of magnetism is generally unimportant in determining the overall magnetic properties of a rock.

An unpaired electron spin behaves like a magnetic dipole and can align with an applied magnetic field, overwhelming the diamagnetic effect. If the unpaired spins are free and do not interact with other spins, a type of magnetism called *paramagnetism* results. It is characterized by a weak alignment of spins parallel to the magnetic field; that is, the paramagnetic susceptibility is positive but weak. Paramagnetic behavior is common in many minerals and rocks in the Earth's crust, and consequently, the ambient geomagnetic field can induce a magnetization in such rocks. The induced magnetization has the same direction as the field except in rocks that are anisotropic—for example, if they have a layered structure.

In a metal or mineral, the alignment of each atomic spin with a magnetic field is also affected by the magnetic fields of neighboring atoms. The relationship between the atoms in a solid therefore determines the type of magnetism it exhibits. The atoms in a metal are packed together to form a regular lattice. In a few metals—for example, iron, nickel, and cobalt—the atoms are so closely spaced that electrons can be exchanged between adjacent atoms. The *exchange energy* associated with this behavior couples the magnetic moments of the atoms tightly to each other and produces a strong internal magnetic field. Often referred to as a molecular field, it binds neighboring magnetic spins in a strong alignment (Fig. 5.1a). This is responsible for the strong magnetic properties of iron, nickel, and cobalt. This type of magnetism is called *ferromagnetism*.

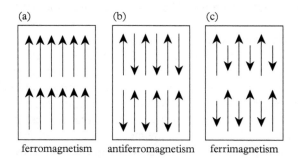

Fig. 5.1 *(a) The parallel alignment of magnetic spins in a ferromagnetic metal produces a strong intrinsic magnetization. (b) The antiparallel alignments of alternating equal spins in an antiferromagnetic crystal result in zero net intrinsic magnetization. (c) The antiparallel alignments of alternating unequal magnetic spins in a ferrimagnetic crystal produce a net magnetization and magnetic properties like those of a ferromagnetic metal.*

The susceptibility of a ferromagnetic material is positive and far larger than a paramagnetic susceptibility. The ferromagnetic magnetization decreases with increasing temperature and is destroyed upon heating above a particular temperature, known as its Curie temperature. This is because the molecular field disappears above this temperature and the strong internal alignment of spins breaks down. As a result, above the Curie temperature the ferromagnetic behavior becomes paramagnetic. An important characteristic of ferromagnetic materials is their ability to become permanently magnetized. When a ferromagnetic material is exposed to a magnetic field, a magnetization is induced in it, but it does not disappear after the field is removed. Part of the induced magnetization is retained permanently and is called a *remanent magnetization*.

5.4   Antiferromagnetic and Ferrimagnetic Minerals

The internal structure of a mineral is quite different from that of a metal in which the atoms are closely packed. Instead, minerals have a crystalline structure formed by closely packed ions (i.e., electrically charged atoms) of oxygen or sulphur. For example, the internal structure of the iron oxides consists of regularly spaced oxygen ions. There are open spaces between the ions into which iron ions can fit because they are smaller than the oxygen ions. However, the spacing between the iron ions then becomes too large for them to exchange electrons directly as in a metal. An *indirect exchange* of electrons between iron ions can take place by way of an intervening oxygen ion. The electron exchange is governed by the rules of quantum physics, which require that the magnetic spins of the exchanged electrons have opposing directions (Fig. 5.1b). A molecular field is produced in the crystal, which causes the magnetic spins of neighboring iron ions to orient antiparallel to each other. If the iron ions have the same number of electron spins as their neighbors, the magnetic moments of the opposing spins are equal and opposite and cancel each other out. The mineral has no net magnetic moment and is classified as *antiferromagnetic*.

However, a special type of magnetic behavior that is akin to ferromagnetism is observed in a few magnetic minerals, especially the oxides of iron such as magnetite, maghemite, and hematite. In these minerals, the magnetic moments of oppositely oriented spins are unequal, and the imbalance gives rise to a strong internal molecular field that results in a net alignment of the atomic magnetic spins (Fig. 5.1c). This type of magnetism is called *ferrimagnetism,* and the materials that possess it are called *ferrites.* At room temperature they have a strong magnetization. In the same way as in a ferromagnetic metal, the ordered internal alignment of magnetic spins is overcome by thermal disorder at a certain temperature, above which a ferrite becomes paramagnetic. The temperature at which this occurs in an antiferromagnet or a ferrite is formally called the *Néel temperature,* after Louis Néel, the French physicist who discovered antiferromagnetism and ferrimagnetism. However, it is often referred to loosely as the Curie temperature.

The common iron oxide hematite has an antiferromagnetic structure, but it behaves like a ferrimagnetic mineral because of imperfections in its lattice; it has a high Néel temperature of 675°C. In some rocks, iron sulfides with defects in their structure (e.g., pyrrhotite) are also important carriers of magnetism. Ferrimagnetic minerals have a high susceptibility compared to the host minerals that make up the bulk of a rock, and the Earth's magnetic field can induce a comparatively strong magnetization in them.

Magnetite and hematite often have titanium in their structures and form solid-solution series of minerals called titanomagnetite and titanohematite, respectively. Titanomagnetite is particularly important. It is the ferrimagnetic mineral in the basalts that form the uppermost layer of the oceanic crust and is responsible for marine magnetic anomalies. The amount of titanium in titanomagnetite strongly influences its magnetic properties. For example, pure magnetite contains no titanium and has a Néel temperature of 580°C, but the titanium content of the titanomagnetite in oceanic basalts reduces their Néel temperatures to 200–400°C.

The magnetization of a ferrimagnetic grain is strongly influenced by its size. A uniformly magnetized grain smaller than one-tenth of a micron (i.e., one-tenth of a millionth of a meter) in diameter can exist as a single-domain (SD) grain. SD grains have strong, stable remanent magnetizations, making it possible for rocks to carry a record of the field in which they formed. However, if a grain grows larger than a critical size, its magnetic energy becomes unstable and the magnetization subdivides to form a multidomain (MD) grain that consists of several smaller domains. Neighboring domains are separated from one another by thin "walls" that are easily moved by magnetic fields. As a result, the remanent magnetization of MD grains is weak and unstable. The mobility of domain walls in a magnetic field makes it easy to induce a magnetization; consequently, MD grains have a higher susceptibility than SD grains of the same magnetic mineral.

The ferrimagnetic minerals, when present in significant amounts, determine the susceptibility and other magnetic properties of a rock. The concentration of the ferrimagnetic minerals is variable and causes the susceptibilities of different rock types to span a very wide range of values, covering more than four orders of magnitude (Fig. 5.2). The susceptibilities of igneous rocks are much higher than those of sedimentary rocks: basalt has a susceptibility 1,000 times higher than limestone. This difference also exists for the remanent magnetizations carried by the rocks.

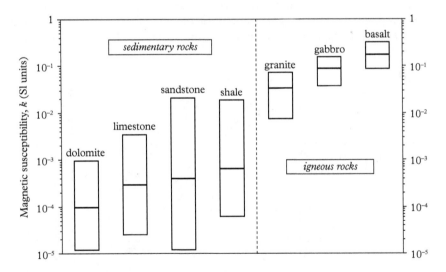

Fig. 5.2 *The ranges of magnetic susceptibility in common sedimentary and igneous rocks, represented by the vertical limits of each box. The horizontal bar in each box is the median value.* (Redrawn after *Fig. 11.19a in W. Lowrie and A. Fichtner,* Fundamentals of Geophysics, *3rd ed., Cambridge University Press, 2020. Reprinted with permission)*

5.5   Induced and Remanent Magnetizations

The magnetization of a rock usually has two components (Fig. 5.3). One component is induced by the present magnetic field, whereas the other is an ancient component that was acquired when the rock was formed. The magnetization $\mathbf{M}_i$ induced by the present Earth's field is proportional to the magnetic susceptibility of the rock, which depends on the concentrations and types of the magnetic minerals it contains. All the minerals in a rock—diamagnetic, paramagnetic, and ferrimagnetic—contribute to the induced magnetization. Its relationship to the magnetic field is expressed by the equation

$$\mathbf{M}_i = k\mathbf{B}/\mu_0 \tag{5.1}$$

in which k is the rock susceptibility and μ_0 is the magnetic constant.

The second type of magnetization in a rock is its remanent magnetization, $\mathbf{M}_r$, which is carried exclusively by the ferrimagnetic minerals in the rock. They make up only a small fraction of any rock, most of which consists of paramagnetic or diamagnetic minerals, but they are usually responsible for its magnetic properties. In some strongly magnetic igneous rocks, the ferrimagnetic fraction can amount to as much as 1% of the rock, but in a sedimentary rock the concentration may be less than one hundredth of a percent (0.01%). The remanent magnetization $\mathbf{M}_r$ is not caused by the present field but dates from an earlier time, when the rock was formed or altered.

The total magnetization of the near-surface rocks is the sum of the induced and remanent components. This is a simple sum of the intensities if the two components are parallel. However, their directions may be quite different (Fig. 5.3a). The induced

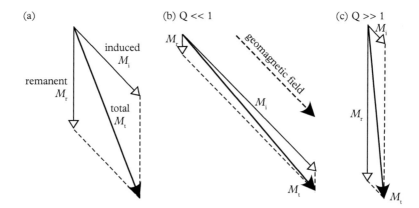

Fig. 5.3 *The relationship between the remanent (M_r), induced (M_i), and resulting total (M_t) magnetizations in a rock. (a) When neither component of magnetization dominates, the direction of the net magnetization M_t lies between M_i and M_r; (b) if M_i » M_r, the resultant magnetization, M_t is close to M_i; (c) conversely, when M_r » M_i, the resultant magnetization M_t is close to M_r.*

component is parallel to the present field, but the remanent component may be in a different direction, corresponding to the direction of the ancient field in which the remanent magnetization was acquired (Fig. 5.3). The total magnetization is found by calculating the *vector sum* of the two components, which takes account of their possibly different directions:

$$\mathbf{M_t} = \mathbf{M_i} + \mathbf{M_r} \tag{5.2}$$

The ratio of the intensity of the remanent magnetization to that of the induced component, M_r/M_i, is called the *Königsberger ratio*, Q. Except in volcanic areas characterized by fine-grained lavas, most of the near-surface crustal rocks that cover the continents have magnetic properties carried by magnetic grains in which the remanent magnetization $\mathbf{M_r}$ is much weaker than the induced magnetization $\mathbf{M_i}$. In this case, the Q factor is very small (Fig. 5.3b, $Q \ll 1$), and the remanent component can often be ignored, so that it is often assumed that the total field is parallel to the inducing field at the site.

The opposite situation is encountered over oceanic crust. Marine magnetic anomalies are due to strongly magnetized basaltic rocks in the upper layer of the oceanic crust. The remanent magnetization $\mathbf{M_r}$ of oceanic basalts is much stronger than the magnetization $\mathbf{M_i}$ induced by the present field and the Q factor is typically 100 or greater (Fig. 5.3c, $Q \gg 1$). The interpretation of oceanic magnetic anomalies is based on the remanent magnetizations in the basaltic layer, and the induced component is neglected.

5.6 The Thickness of the Magnetized Crustal Layer

The magnetic susceptibility of a ferrimagnetic mineral decreases with increasing temperature. Above its Néel temperature, the mineral is paramagnetic and its susceptibility is negligibly small. Locally, the depth at which the Néel temperatures is reached determines the thickness of the magnetized layer of crustal rocks. In continental crust the temperature increases with depth at a mean rate called the *geothermal gradient*, which is typically around 20–30 °C/km but depends on local geological conditions. The maximum Néel temperatures in continental crustal rocks are 580 °C if the ferrimagnetic mineral is magnetite and 670 °C if hematite predominates. Beneath the continents, these temperatures are reached at depths of 20–35 km. The thickness of the continental crust varies between a common value of 30–40 km and as much as 70 km under a mountain range. However, the geothermal gradient decreases with depth, which causes the Néel temperature to occur at greater depth. Thus, where the crust is thinnest, the magnetized continental layer may include part of the underlying upper mantle as well.

The oceanic crust is much thinner than the continental crust. Formed by seafloor spreading at a divergent plate boundary, oceanic crust has an average thickness of around 7 km. The thickness is variable, as it depends on the rate of seafloor spreading at a ridge. It is thinner (~ 5 km) for crust formed at slowly spreading ridges and thicker (~10 km) for fast-spreading ridges. The magnetic mineral in the oceanic basalt that forms the top igneous layer of oceanic crust is titanomagnetite and has a Néel temperature of 200–400°C. The vertical temperature gradient in the deep ocean basins averages about 45°C/km but can be 40–80°C/km in volcanic arcs. Thus, the Néel temperature is reached at shallower depth beneath the oceans than under the continents. However, the layer of gabbroic rock beneath the basalts contains magnetite with a higher Néel temperature of 580 °C, and it probably also contributes to the crustal magnetization. The magnetic structure of the oceanic crust is discussed in more detail in Chapter 5.10.

5.7   How a Magnetic Anomaly Originates

A magnetic anomaly originates when rocks with different magnetizations are in contact with each other. For example, it may arise where a strongly magnetic basaltic dike intrudes into a weakly magnetic granitic pluton or where folding or faulting has brought rock types with contrasting magnetizations into juxtaposition. Magnetic anomalies are not found over crust that is uniformly magnetized or that is free of tectonic features.

   Magnetic anomalies in igneous rocks are of direct interest in exploration geophysics. They may identify a local intrusion that is enriched in valuable metals or minerals, which could make it a suitable target for commercial exploitation. Some magnetic anomalies are of regional extent as a consequence of large-scale tectonism. In order to obtain an overview of a country's subsurface geology or its potential mineral assets, airborne magnetic surveys of the country are often carried out by a governmental organization, such as a Geological Survey.

   The discontinuity in magnetization across a surface of contact between different rocks produces secondary magnetic fields, which combine to form a magnetic anomaly. Its shape is determined by several factors, including the geographical location and shape of the body. The amplitude of the anomaly depends on the body's depth below the measurement surface, and on the size of the magnetization contrast, which, in turn, reflects the differences in susceptibility or remanent magnetization between the body and adjacent rocks.

   Instead of plotting individual measurements at the locations where they are made, the data may be represented on a map by contour lines (Fig. 5.4a). Smooth curves connect places where the magnetic field has the same value (i.e., declination, inclination, or intensity). The positions of the contour are interpolated where they pass between the places where the measurements were made. The contour lines of an anomaly isolate regions where the measured field deviates from the regional trend. A profile across the local anomaly (Fig. 5.4b) emphasizes its shape and is used in interpretation of the size and depth of the structure responsible for the anomaly.

   A simple example illustrates how a magnetic anomaly results from the magnetization induced by the present field in a subsurface body. A body that has a roughly equal extent horizontally and vertically can be modeled to a first approximation by a sphere

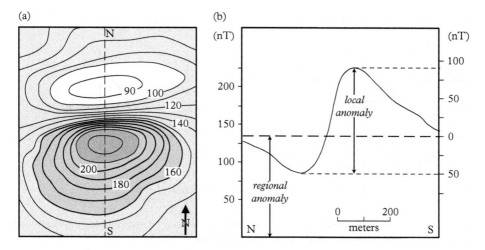

Fig. 5.4 *(a) Contour lines of magnetic intensity (in* nT*) over an anomalously magnetized structure; (b) interpretation profile along the line from N to S across distinctive features of the anomaly.*

(Fig. 5.5a). Assume that the inducing field is vertical. In reality, this would only be the case at the magnetic dip poles, but a mathematical technique (called rotation to the pole) actually allows manipulation of the measured anomaly to achieve this effect. Assume further that the body has a high susceptibility, so that any remanent magnetization is negligible ($Q \ll 1$). The vertical magnetizing field induces a vertical magnetization in the body. Its anomaly field can be represented to a first approximation as that of a dipole.

At positions A and G on a profile across the body at the surface of the ground, the direction of this secondary field is slightly upward, opposite to the inducing field and thereby forming a negative anomaly (Fig. 5.5b). However, A and G are at the ends of the profile, where the magnetometer is far from the magnetized body, so the negative anomaly is very small. With progress along the profile toward the axis of the body, the anomaly is more noticeably negative. At positions B and F, the secondary field is horizontal; it acts at right angles to the vertical inducing field and therefore causes zero anomaly. At positions C and E closer to the axis of the body, the anomalous field acts downward and reinforces the inducing field, creating a positive anomaly, which reaches a maximum at position D over the axis of the body.

The anomaly in this idealized example is symmetrical about the axis of the vertically magnetized structure. In most cases, the magnetizing field or the shape of the structure would be inclined rather than vertical. An inclined field or structure causes the anomaly to be asymmetrically shaped. For example, if the magnetizing field in Fig. 5.5a were inclined from top right to bottom left, the induced magnetization of the body would align accordingly, changing the shape of the anomaly. The central peak would shift to the right of the axis, the left-side lobe would become more negative, and the right lobe would decrease (Fig. 5.4b) and perhaps disappear. In practice, magnetic anomalies

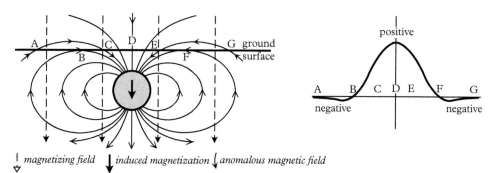

Fig. 5.5 *(a) Sketch of the anomalous magnetic field lines of a vertically magnetized body; (b) variation of the amplitude of the anomaly along a horizontal profile over the body.*

are not perfectly shaped, and sophisticated data processing is an important factor in postsurveying evaluation.

5.8   Continental Magnetic Anomalies

The continental "shield" areas, where the oldest rocks occur, have experienced numerous episodes of tectonic deformation, involving reheating and recrystallization of their minerals. Moreover, the shield areas in polar latitudes have experienced Ice Ages, during which glaciers several kilometers thick scraped away the surface of the land, exposing coarse-grained plutonic rocks that were formed deep in the crust. The magnetizations of plutonic rocks have low Königsberger ratios because the grain sizes of their magnetic minerals are multidomain, and they are easily magnetized in the present field. Consequently, continental magnetic anomalies are induced mainly by the present geomagnetic field, and remanent magnetizations in the rocks may be relatively unimportant.

Several prominent magnetic anomalies in the continental crust are due to particularly strong enrichment of metallic orebodies. One of the world's largest magnetic anomalies covers an area 800 km long and 200 km wide in the Kursk region of southwestern Russia, adjacent to the border with Ukraine. The *Kursk anomaly* is due to extensive lodes of iron ore, which produce regional anomalies with amplitudes of several 1,000 nT at Earth's surface. At satellite altitude in a low-Earth orbit, the Kursk anomaly is still prominent and measures around 35 nT. Another large anomaly is found on the magnetic equator in the Central African Republic. Named the *Bangui anomaly*, it measures about 1,000 km in length east to west and 700 km in width from north to south. The anomaly is strongly negative and has an amplitude of around 1,000 nT at the surface and 20 nT at MAGSAT altitude. The cause of the anomaly is controversial. According to one theory, it is due to a large intrusion of magma at depth, but an alternative theory posits that it is the site of a meteoritic impact in Precambrian time.

5.9 The Magnetization of the Oceanic Crust

The oceanic crust has a layered structure. The sediments that cover the oceanic depths are denoted oceanic Layer 1. This layer is thin to nonexistent on the ridge axis and thickens to several hundred meters with increasing distance from the ridge. The sediments contain fine-grained magnetite but in such weak concentrations (often less than 0.01%) that the sediments are only weakly magnetic. Their magnetizations can only be measured with special magnetometers under laboratory conditions. As a result, magnetic surveys over oceanic territories treat the sediments as if they are nonmagnetic and transparent to magnetic fields. The important magnetization of the oceanic crust is attributed to the igneous rocks beneath the sediments.

From a geological point of view, however, the sediments carry an important magnetic record related to their deposition. During and after deposition, but before consolidation of the sediment, the magnetite grains acquire an alignment with the ambient magnetic field. The field has changed polarity on average four to five times per million years in the past 20 million years. However, the reversals occur at very irregular intervals, with the most recent full reversal happening 780,000 years ago. The vertical sequence of remanent magnetizations of magnetite in the deep-sea sediments preserves the polarity record, which becomes conserved in the sediment during diagenesis. Millions of years later, it provides geologists with a record of geomagnetic field polarity during deposition of the sediment. This record correlates with the horizontal record of polarity acquired by the igneous oceanic crust during seafloor spreading.

The top layer of the igneous basement is about 500 m thick and consists of basaltic lavas that were extruded at the ridge, forming oceanic Layer 2A. The next layer beneath the lavas, oceanic Layer 2B, is about 1 km thick and consists of basaltic rocks that did not reach the surface but were intruded as dikes into the overlying layer. As a result, Layer 2B cooled more slowly than the lavas and the grain size of the basalt is coarser.

Beneath the basalts lies a layer of gabbroic rocks, some 4–5 km thick (oceanic Layer 3). They are coarse-grained, with a composition similar to (but more variable than) the basaltic layer. They have lower Königsberger ratios ($Q \ll 1$) and acquire weaker, less stable thermoremanent magnetizations than Layer 2 rocks. The depth of the layer increases the distance from the surface magnetometers, and the higher temperature weakens the magnetization, so Layer 3 does not contribute significantly to the measured oceanic magnetic anomalies.

The upper mantle beneath continental and oceanic crust is made up of coarse-grained igneous rocks called peridotite, in which the primary mineral is olivine. They are observed on the surface in ophiolites at some destructive margins of tectonic plates, where slabs of oceanic lithosphere have been overthrust onto continental crust instead of being subducted beneath it.

During the Second World War, scientific equipment was developed for military purposes that would later prove to be invaluable for geophysical exploration of the world's oceans. Sonar, a method of echo-sounding, was invented for use in submarine warfare and adapted by hydrographers to map the depths of the oceans. Away from a shallow shelf that abuts each continent the oceans are several kilometers deep. Along the borders of the Pacific Ocean there are extremely deep, narrow trenches; the deepest, the Marianas trench, extends to around 11 km in depth. Extensive surveying of the ocean basins by oceanographic institutes has revealed submarine mountain chains, several thousand kilometers in length, that rise thousands of meters above the ocean floor. Designated as oceanic ridges, they often have a "graben"—a steep-sided, ditch-like central valley—along their crests.

Most of the world's large earthquakes occur near the deep oceanic trenches, and many of them have focal depths of several hundred kilometers. The distribution of their locations and depths defines planes of seismic activity called Benioff zones that dip beneath the adjacent continents. In the 1950s, seismologists discovered that the oceanic ridges are also seismically active, although the earthquakes on the ridges are not as large as those at Benioff zones. Seismic analysis of the first motions of the ground during these earthquakes has revealed that opposite sides of the ridges are moving away from each other. Their rates of separation are only 20–120 mm/yr, which is very slow in human terms but fast on a geological time scale. The process of separation at the ridge is called seafloor spreading.

In the 1950s, marine surveys began to employ total-field magnetometers to measure the magnetic field over the oceans (Chapter 2). Conspicuous patterns of lineated anomalies with amplitudes of several hundreds of nanotesla were discovered in the eastern Pacific Ocean off the west coast of the United States and Canada. The patterns were found to be associated with oceanic ridges that had been offset by great horizontal faults. The directions of motion on these oceanic faults were observed to differ from those on horizontal continental faults; they were named transform faults. The pattern of strong magnetic anomalies, symmetric across the axis of the ridge and elongated parallel to its axis, was found at other ridges in the Antarctic, South Atlantic, and North Atlantic Oceans.

The conventional method for interpreting anomalies on the continents assumed that magnetic anomalies were due to contrasts between magnetizations induced by the present field in rock units with different susceptibilities. However, when this method was applied to the oceanic anomalies, it failed to explain the striped anomaly pattern. A solution to the dilemma was found by attributing the oceanic magnetic anomalies to remanent magnetizations acquired in ancient magnetic fields, rather than to magnetizations induced by the present field (Vine & Matthews, 1963). The hypothesis made it possible to relate the symmetric pattern of anomalies at oceanic ridges to the seafloor spreading that was indicated by the ridge seismicity. The key to the mechanism (Fig. 5.6) can be explained as follows.

When basaltic magma erupts at an oceanic ridge to form oceanic Layer 2A, it is quenched (i.e., it cools very rapidly) on contact with the cold seawater, forming a very fine-grained rock. While it cools below the Néel temperature, the basalt acquires a thermal remanent magnetization (TRM) in the direction of the ambient magnetic field (see also Chapter 6.1). Long linear magnetic anomalies are formed parallel to the strike of the ridge above stripes of magnetized crust. The magnetizations beneath adjacent stripes have opposite polarities, alternating between "normal" (i.e., the same as present-day)

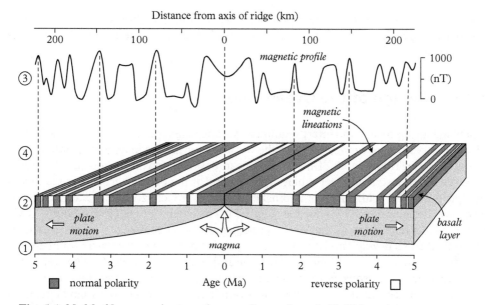

Fig. 5.6 *Model of how oceanic magnetic anomalies are formed. (1) Rising magma erupts constantly at an oceanic ridge, spreads laterally, and becomes magnetized while cooling in the ambient field. (2) Irregular reversals of magnetic field polarity during sea-floor spreading at the ridge are recorded as alternately magnetized blocks of oceanic crust on both sides of the ridge. (3) A magnetic survey across the ridge measures positive and negative anomalies over (4) alternating blocks of normally and reversely magnetized crust.*

and "reverse." The large contrast in TRM intensity at the contact between oppositely magnetized stripes accounts for the large amplitudes of oceanic magnetic anomalies. The symmetric pattern of anomalies documents the irregular changes in geomagnetic polarity during seafloor spreading. The oceanic magnetization is akin to a tape recording of the history of changing geomagnetic field polarity. This discovery verified the concept of seafloor spreading and revolutionized geological understanding of global tectonics.

In the late 1960s, paleomagnetic analyses of radiometrically dated continental lavas revealed a record of irregularly spaced magnetic polarity reversals during the past 5 million years. A sequence of magnetic reversals was soon established in marine sediments cored from the deep ocean basins. The reversal sequence deduced from seafloor spreading at an oceanic ridge was found to be identical to the sequence in continental lavas and deep-sea sediments. By correlating the marine magnetic anomalies to the dated reversal history, it was possible to compute the rate of seafloor spreading at a ridge. The ages of older anomalies on long magnetic profiles were estimated by extrapolation, assuming that seafloor spreading at a particular ridge takes place at a constant rate. This is a good assumption but the correlations between very long profiles in different ocean basins showed that spreading rates are not absolutely constant at any given ridge. Later, the development of magnetic stratigraphy made it possible to correlate the marine polarity record with dated stratigraphic sections on land. By this means, the history of polarity changes of the magnetic field was extended back in time through most of the past 200 Myr.

5.10   The Age of the Ocean Floor

Following the discovery and interpretation of marine magnetic anomalies, extensive geophysical surveys were carried out over the world's oceans. The magnetic lineations were found at ocean ridges in all of the major ocean basins. Later, a second set of lineated anomalies was described, over much older regions of the oceanic crust. For identification purposes, the younger set of anomalies, found over Cenozoic and Late Cretaceous crust, was designated the C-sequence (Fig. 5.7). The older set (aged from about 170 Ma to 126 Ma) is referred to as the Mesozoic sequence (or M-sequence). In order to identify individual Cenozoic and late Cretaceous marine anomalies, they were numbered in increasing order away from their oceanic ridge and assigned the prefix C. They represent intervals of time, or chrons, in which geomagnetic polarity was constant. The normal and reverse polarity chrons that correspond to a numbered anomaly were assigned the label "n" or "r," respectively. Using this nomenclature, the Cenozoic sequence is numbered from the youngest polarity chron C1n at an ocean ridge to reverse polarity chron C33r in the Late Cretaceous. Chron C33r is preceded by a long interval of normally magnetized crust, which was originally labeled polarity chron C34n.

The region of oceanic crust corresponding to chron C34n is known as the Cretaceous Quiet Zone. It was formed during a time when the geomagnetic field maintained a constant normal polarity for 35–40 My, which is far longer than any other Cenozoic polarity chron. In acknowledgment of its exceptional duration, the corresponding time interval is usually called the Cretaceous Normal Polarity Superchron, replacing the earlier C34n label.

The normal polarity superchron was preceded by a negative anomaly labeled M0r, which has an estimated age of 126 Ma; the normal polarity anomaly that precedes it is called M1n. The M-sequence anomalies are then numbered in order of increasing age. The oldest resolvable by sea-surface magnetic surveys is anomaly M25r (Fig. 5.7), with an estimated Late Jurassic age of close to 156 Ma. The oceanic crust older than the M-sequence is characterized by a "Jurassic Quiet Zone." However, it differs in nature from the Cretaceous Quiet Zone. It appears not to represent a time when the field had constant polarity. Investigation of this quiet zone is technically demanding and still incomplete.

The Jurassic Quiet Zone lies over the oldest oceanic crust, in the northwest Pacific Ocean, where the ocean basins are extremely deep. The depth of the seafloor locally can be greater than 6 km, whereas other ocean basins are about 3–4 km deep. The increased distance of the magnetometer from the magnetized layer weakens the measured magnetic anomalies. To obtain a better resolution, the Jurassic Quiet Zone has been surveyed with deep-towed magnetometers. In this procedure, the instruments are towed behind a research vessel at a height of typically 60–100 m above the seafloor. The M-sequence of anomalies was extended by this means to anomaly M44, which has an estimated age of around 170 Ma. The amplitudes of magnetic anomalies that are even older are too weak to be securely correlated as lineations.

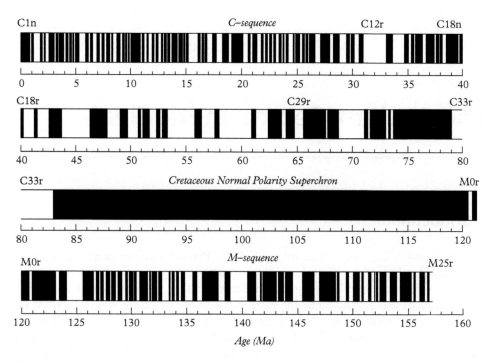

Fig. 5.7 *The sequences of magnetic polarity chrons recorded in the magnetization of the oceanic crust. Subsequent revised dating associates slightly different ages with the reversals than those indicated. (Redrawn after: Fig. 1 in W. Lowrie and D. V. Kent, Geomagnetic Polarity Timescales and Reversal Frequency Regimes. In Timescales of the Paleomagnetic Field, Geophysical Monograph Series 145, editors: J. E. T. Channell, D. V. Kent, W. Lowrie, and J. G. Meert. American Geophysical Union, 2004)*

The ages of marine magnetic lineations make it possible to date correlative geological processes, such as the rates of motion of global tectonic plates and the timing of geological events that result from interactions at plate margins. A direct way of ascertaining the ages of marine magnetic anomalies is to drill deep holes through the sedimentary layer into the underlying basaltic lava and to recover cores of rock from the drillhole. Throughout the past 60 years, international consortiums—known by the acronyms JOIDES, DSDP, ODP, IODP—have organized a continuous series of expeditions with dedicated drilling vessels to achieve this goal. In the initial years, cores were obtained only from the sedimentary layers, although fragments of the igneous layer were also recovered. Later expeditions were able to drill deep into the basaltic basement. The projects resulted in the drilling of hundreds of drillholes into the oceanic crust in all of the world's oceans. For example, in the first series of marine drilling expeditions by the research vessel *Glomar Challenger*, holes were drilled at 624 sites in water depths of up to 7 km, penetrating the crustal rocks for more than a kilometer, even into the basaltic layer. The ages of basal sediments in the cores are found to increase with distance away from the ridge, consistent with the hypothesis of seafloor spreading.

6

The Ancient Geomagnetic Field

Introduction

In the late 1940s, the Nobel Prize winner P. M. S. Blackett became interested in understanding the origin of the Earth's magnetic field. He theorized that it was related to the Earth's rotation, and he designed an experiment to test the idea. The result of the experiment was negative but the instrument he designed for the test—an astatic magnetometer—would be used to revolutionize geological knowledge by providing clues to the mobility of continents. The device consisted of a pair of small horizontal magnets about 15 cm apart, mounted one above the other on a vertical rod, so that they pointed in opposite directions. The assembly was suspended from a thin quartz fiber. In a uniform magnetic field, the magnets experienced opposite torques, and the assembly did not move. But if a magnetized sample was brought near to one of them, the magnetic field of the sample disturbed the balance. The resulting torque caused the assembly to rotate, twisting the fiber until its tension balanced the magnetic torque. The amount of rotation was calibrated in terms of the strength of the magnetization. The device was so sensitive that it could measure the weak remanent magnetization of rocks. It thus contributed to understanding how igneous and sedimentary rocks can become permanently magnetized in the weak geomagnetic field. This made possible a new branch of research into the directions and strength of the geomagnetic field in the geological past.

The astatic magnetometer has long been superseded by even more sensitive instruments designed around quantum mechanical principles, but its invention in 1952 opened the door for pioneering analyses of the magnetization of rocks and the magnetic minerals in them. Geologists used it to study the magnetization of crustal rocks, first in Great Britain and North America. The early paleomagnetic data were poorly defined by modern standards, but they provided strong evidence for "continental drift" between North America and Europe. Within a few years paleomagnetic investigations were being carried out on all the continents.

Subsequently, paleomagnetic research has enabled scientists to understand the slow relative movements between all the continents during hundreds of millions of years and to reconstruct the assembly and breakup of former supercontinents. By the late 1960s, it had also been established that the geomagnetic field has reversed polarity several times at irregular intervals during the past four to five million years. Paleomagnetism provided the

The Earth's Magnetic Field. William Lowrie, Oxford University Press. © William Lowrie (2023).
DOI: 10.1093/oso/9780192862679.003.0006

means of dating the lineated magnetic anomalies in the oceanic crust that are associated with seafloor spreading at oceanic ridges. Gradually, a record of hundreds of polarity reversals has been established that stretches far back into the history of the planet and has been dated by correlation to the geological time scale.

6.1 The Natural Remanent Magnetizations of Rocks

Approximately two-thirds of the rocks that crop out on the Earth's surface have a sedimentary origin. The remaining third consists of crystalline rocks, divided in roughly equal measure between igneous and metamorphic types. The igneous rocks, which form when magma from the interior reaches the crust, are again divided almost equally between intrusive and extrusive types. Intrusive rocks (e.g., dikes, plutons) form within the crust, whereas extrusive rocks (e.g., lavas) are erupted and form on top of preexisting crust.

It is perhaps odd to think of rocks as magnets, but they are indeed magnetic. As described earlier (Chapter 1.1), magnetism was originally discovered as a property of a special type of stone called lodestone. However, a rock's magnetism is usually very weak and can only be measured with special, dedicated magnetometers. Rocks become magnetized when they form in the Earth's magnetic field, and their magnetism can change if they are later heated or altered. As a result, the natural remanent magnetizations (NRM) of rocks often consist of multiple components acquired at different times in the rock's history. Paleomagnetic specialists have developed laboratory techniques to define and describe the original component of NRM. It is acquired by different mechanisms in igneous and sedimentary rocks.

Igneous Rocks

An igneous rock becomes magnetized in the Earth's magnetic field at a very high temperature, after the crystallization of its minerals has been completed. The temperature at which molten lava solidifies is often higher than $1,000°C$, which is well above the Néel point of any ferrimagnetic mineral the rock may contain (Fig. 6.1a). On cooling below its melting temperature, it solidifies, and the positions and orientations of its mineral grains become fixed. However, the elementary magnetic moments (Chapter 5.2) inside the ferrimagnetic grains are still free to change direction and are subject to thermal fluctuations. The magnetic mineral is paramagnetic at this high temperature (Fig. 6.1b).

On subsequent cooling, the temperature of a grain of magnetite or hematite sinks below its Néel point ($580°C$ and $675°C$, respectively) and the grain becomes ferrimagnetic. The magnetic moments within the grains now align as closely as possible with the geomagnetic field. (It is important to understand that only the magnetism can align with the field: the grains themselves are fixed in the solid mineral.) The alignment of magnetic moments is not perfect, but there is a statistical bias to align close to the field

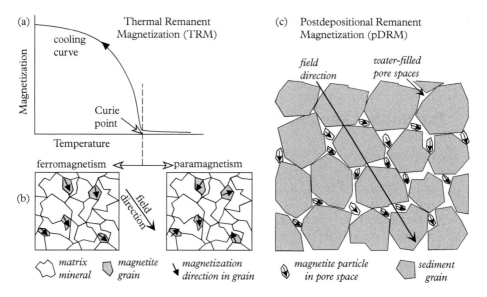

Fig. 6.1 *(a) Acquisition of thermal remanent magnetization (TRM) when an igneous rock cools through the Néel temperature of its ferrimagnetic minerals. (b) The directions of the magnetic moments of ferrimagnetic grains, which are random above the Néel temperature (right) and partially aligned after cooling (left). (c) Acquisition of postdepositional remanent magnetization (pDRM) in sediments. (Redrawn after Figs. 12.07a and 12.09a in W. Lowrie and A. Fichtner, Fundamentals of Geophysics, 3rd ed., Cambridge University Press, 2020. Reprinted with permission)*

direction. Shortly after the temperature of the rock drops beneath the Néel temperature of the magnetic mineral, the magnetic alignment becomes "blocked" and is immovable during further cooling. The mineral has acquired a remanent magnetization, which on cooling to ambient temperature is effectively locked into the rock.

The magnetization acquired in this way, when an igneous rock cools through its Néel point, is called a thermal remanent (or thermoremanent) magnetization (TRM). If the grain size of the rock is fine enough for the magnetic mineral to have single-domain size (Chapter 5.2), a TRM is a very stable magnetization. In the absence of later reheating or alteration, it can preserve the direction of the magnetic field in which it was acquired for millions of years. The TRMs in continental and oceanic basalts have preserved the ancient record of magnetic field polarity for hundreds of millions of years. Some ferrimagnetic minerals in Precambrian rocks still possess an original TRM that is billions of years old.

Sedimentary Rocks

Mountain ranges like the Apennines and Alps are formed by horizontal forces that fold and crumple the Earth's crust. Subsequently, glaciers and rivers slowly erode the mountains. The material eroded from rock outcrops on land by physical abrasion or chemical dissolution is transported into lakes and seas, where it is deposited as an unconsolidated sediment. A small proportion of the detritus is often magnetite, and occurs in various grain sizes. The rate at which the grains settle during deposition depends on their size. The coarsest grains are deposited first, closest to the source, while the finest grains remain suspended longer in the water before settling and may be deposited far from land. In the deep seas and oceans, the sedimentary deposits consist of the chalky shells of tiny planktonic marine animals called foraminifera that populate the marine world.

Some magnetic minerals originate in the oceans. They derive from certain types of marine bacteria classified as *magnetotactic* because they react to a magnetic field. The magnetotactic bacteria produce tiny magnetite crystals that link together to form chains, called *magnetosomes*. They are the most important source of magnetite in pelagic limestones, which is a type of limestone deposited far from land. As a result of the physical and chemical processes during deposition, marine limestones like those in the Apennines contain a minute assemblage of submicron-sized grains of magnetite, often amounting to less than 0.01% of the rock by volume.

Very small magnetite particles take a very long time to settle, even in still water. The action of the Earth's magnetic field orients the particles statistically during settling. The orientation is partially lost through mechanical rotation when the particles touch bottom, in the same way that the shape of a dropped book causes it to rotate and fall flat on contact with the floor. The sediment cannot preserve perfectly the direction of the field that oriented the particles during settling. As a result, the inclination of the magnetization in a sediment may have a "flattening" error.

In some sediments, however, the remanent magnetization is acquired only after deposition. For example, the magnetite crystals formed by magnetotactic bacteria are so small that they can float in the interstitial pore spaces of a sediment, where they align with the ambient magnetic field (Fig. 6.1c). The water is squeezed from the sediment during consolidation of the rock, and the magnetic alignment produced in the depositional field becomes fixed, forming a postdepositional remanent magnetization (pDRM). Although later components of magnetization may be superposed, the pDRM is a stable magnetization that can retain the paleodirection through later tectonic processes, such as folding and faulting. For example, the original directions of magnetization of some of the folded limestones in the Italian Apennines have been preserved for more than a hundred million years.

Metamorphic Rocks

Metamorphic rocks are of little interest from the point of view of preserving a record of the geomagnetic field. The thermal and chemical effects during metamorphism can reset the original magnetizations of rocks, and structural displacements that accompany tectonism can spoil the original directions. Changes in the chemical composition of the magnetic mineral can result in the formation of a new chemical remanent magnetization (CRM), which can partially or completely overprint the original remanent magnetization. A postformation remagnetization can develop in rocks that carry original TRM or pDRM, but laboratory techniques have been developed for identifying the secondary effects and avoiding the use of rocks in which they are pronounced.

6.2   The Geocentric Axial Dipole Hypothesis

Paleomagnetism involves the analysis of the ancient geomagnetic field recorded in the magnetizations of rocks. Its central assumption is the geocentric axial dipole (GAD) hypothesis. This theory states that the time-averaged, geologically significant magnetic field is that of a dipole, located at the center of the Earth and oriented along the rotation axis. The geocentric axial dipole is represented by the Gauss coefficient g_1^0 in the spherical harmonic representation of the field (Chapter 3.5). According to the GAD hypothesis, secular variations of the field ensure that all other Gauss coefficients average to zero over a sufficiently long interval of time, which in practice has to be longer than several millennia. The significance of paleomagnetic data depends on the validity of the GAD model. It appears to be largely valid for at least the past 250 Myr. Although there is strong paleomagnetic evidence for persistent quadrupolar and octupolar components in the time-averaged recent and ancient fields, the assumption of a GAD field still serves as a good working hypothesis.

The inclination and declination of the characteristic remanent magnetization (ChRM) of a rock provide important geological information about the rock's history. They can be used to determine the location of the geomagnetic pole when the rock formed. The calculation must take into account that the present location, and even the continent, where the rock was sampled may have moved since the time of formation. A small error is usually associated with each measurement or group of measurements. The computed pole location is where the magnetic pole *appears* to have been when the rock formed and is called a *virtual* geomagnetic pole (VGP) position.

The VGP position is calculated by using the measured inclination and declination of the magnetization of a rock sample. The distance to the magnetic pole from the site where the magnetization was acquired must first be calculated. The calculation uses the magnetic inclination (I), which depends on the angular distance (θ) of the site from the magnetic pole. The angular distance from the pole and the angle of inclination are related for a GAD field by trigonometric tangent functions:

$$\tan \theta = \frac{2}{\tan I} \tag{6.1}$$

The angular distance θ is the complement of the magnetic latitude λ and is called the magnetic co-latitude (i.e., $\theta = 90 - \lambda$). The declination (D) of the magnetization gives the azimuthal direction, or heading, from the sampling site to the VGP position, which is located at an angular distance θ along that direction. When a sufficient number of VGP positions for rocks of the same age from the same location are combined statistically, the mean position is called a paleomagnetic pole (or, more simply, *paleopole*) and represents the time-averaged field.

The success of paleomagnetism depends ultimately on the validity of the GAD hypothesis. If it is not valid, the significance of apparent polar wander (APW) paths and associated geological reconstructions, as well as the polarity of the ancient geomagnetic field are cast in doubt. The hypothesis is known to be a good, but imperfect, model for the paleomagnetic field. Measurements on rocks and deep-sea sediments that are younger than about 5 Myr have shown that persistent quadrupole and octupole components of the field, each amounting to approximately 2–5% of the strength of the dipole, can survive the time averaging. As a result, deviations from the GAD field are small. Many studies have shown that the GAD hypothesis works well for the paleofield during the Phanerozoic, that is, the past 541 Myr. However, the older rocks are, the greater is the possibility that their original magnetizations have been altered, and the more difficult it becomes to confirm the validity of the hypothesis. There is also the possibility that the ancient field was less dipolar than the present field.

The inclination of a GAD magnetic field is related to the latitude or co-latitude of the site where it is measured (Eq. 6.1). The area of the Earth's surface, at which the field has a given inclination, decreases from the equator to the pole. This affects the frequency of occurrence of a chosen inclination in the paleomagnetic database, which can be predicted for a GAD field as well as for a GAD field with added quadruple and octupole components. In an evaluation of the paleomagnetic database (Kent & Smethurst, 1998), the validity of the GAD hypothesis was tested and found to describe the field well in the Cenozoic and Mesozoic (i.e., younger than about 250 Myr). However, the frequency distribution of inclinations in the Paleozoic and Precambrian did not fit a GAD field perfectly but had an excess of lower inclinations than predicted by the model. This could be accounted for by a field in which axial nondipolar components were much stronger than in the present field. A quadrupole component equivalent to 10% of the axial dipole and an octupole component of 25% gave the optimum fit. However, an alternative interpretation was that the GAD hypothesis is valid and that the results indicate that the paleogeographic distribution of the drifting continents on the early Earth may have had a preference for low latitudes.

6.3 Methods of Paleomagnetism

A typical investigation in an undeformed rock formation with a known geological age consists of three phases: sampling, measurement, and interpretation. Small oriented samples of the rock are taken at a number of suitable outcrops in a geological formation. Laboratory analysis of the direction of the remanent magnetization eliminates unstable components of magnetization that may be present and defines the stable component. Two methods of progressive demagnetization are commonly used: a thermal technique and application of an alternating magnetic field. The thermal demagnetization consists of heating the sample to a chosen temperature to destroy part of the natural magnetization;

subsequently, the sample is cooled in a field-free space to ensure that it does not acquire a new TRM in the laboratory. The procedure is repeated to ever higher temperatures until the NRM is destroyed. Alternating-field demagnetization with stepwise increases in the peak field can also be used to progressively demagnetize the NRM; it is especially useful for samples that would be damaged if heated.

The sequence of partially demagnetized directions is analyzed to derive the direction of the stable component of magnetization acquired when the rock was formed, which might be only a small fraction of the original NRM. The stable component is the ChRM. Field tests have been designed to verify that it is primary, that is, representative of the field present at the time of the original magnetization. The most common test is a fold test in which directions of magnetization are compared around a folded structure. If correcting the directions for the dip of the beds around the fold (effectively "flattening" it) causes the directions to group more tightly, the magnetization is primary. A reversals test consists in comparing normal and reverse directions of magnetization, which should be antipodal if they are primary. A baked contact test can sometimes be used to compare the direction in an igneous rock with the remagnetized direction in a sedimentary rock with which it is in contact. If both rocks have the same direction of magnetization, it is stable and has the age of the igneous rock.

The VGP positions computed from the ChRM directions of rocks of Pleistocene and Pliocene age, that is, younger than 5 Ma, from North America illustrate the dominance of the geocentric axial dipole (Fig. 6.2). Instead of being located around the present magnetic north pole, they are clustered around the Earth's rotation axis, which is the direction of the geocentric axial dipole, in good agreement with the GAD hypothesis.

In order to determine the paleointensity (i.e., strength) of the field in which a rock acquired its remanent magnetization, it is necessary to heat a sample of the rock. This may alter the original magnetic mineralogy and can produce unwanted secondary magnetizations. The most commonly used experimental method was developed by the French scientist E. Thellier in 1937 and later augmented with additional checks to ensure that the rock is not altered during the measurements. The Thellier method and its modifications can be applied to materials such as igneous rocks and pottery, in which a TRM is carried by single-domain-sized grains of magnetic mineral (Chapter 5.2). A rock sample is heated to a chosen temperature, thereby destroying part of the remanent magnetization. It is then cooled to room temperature in a field-free space produced by magnetic shielding, so that it cannot acquire a new TRM as it cools. The sample is reheated to the same temperature as before and is again cooled to room temperature, this time in a known field, so that it acquires a new TRM in the known field. The procedure is repeated stepwise, in each step heating to a progressively increasing temperature. By comparing the amount of the original TRM lost in the field-free heating steps to the TRM gained in the control-field cooling steps, the paleointensity of the field can be computed in terms of the control field.

It is not possible to use the Thellier method for sedimentary rocks or unconsolidated sediments. However, in deep-sea sediments it has been possible to track relative changes in intensity by normalizing the NRM with a suitable proxy for the volume of magnetic mineral in the sample. The susceptibility of the sediment is often used for this purpose.

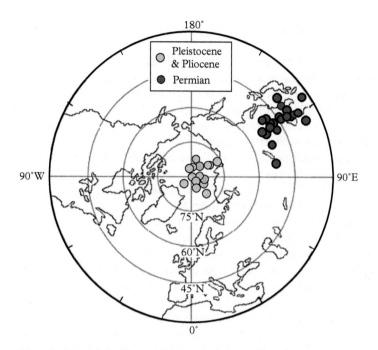

Fig. 6.2 *The locations of North American paleomagnetic poles of Pleistocene and Pliocene age (< 2 Ma) are clustered around the rotation axis,
while older poles of Permian age (> 230 Ma) are clustered over northeast
China. (Redrawn after Fig. 12.20 in W. Lowrie and A. Fichtner,* Fundamentals of Geophysics, *3rd ed., Cambridge University Press, 2020.
Reprinted with permission)*

More appropriately, a remanent magnetization that mimics the coercivity distribution
in the sample is induced as a normalizing factor. The normalized magnetization of the
sediment shows relative changes in intensity of the field during deposition.

6.4   Apparent Polar Wander and Continental Reconstructions

The paleomagnetic poles for rocks of Permian age (i.e., with ages of 250–300 Ma) from
North America cluster around a position in northern China, far from the present location of the geographic axis (Fig. 6.2). This displacement allows two interpretations:
either the magnetic pole—and by implication the rotation axis—was located closer to
the equator in Permian time, or the Earth's crust and lithosphere have moved relative to
the rotation axis. The long-term attitude of the rotation axis is constant, so the second
interpretation is correct. The Permian paleopoles only *appear* to have moved; instead, the
paleomagnetic data document the motion of the lithosphere relative to the rotation axis

as a consequence of plate tectonics. The phenomenon is called apparent polar wander (APW).

Suppose that rocks of the same age from two separate continents give different paleopole locations. By the above argument, moving the continents so that their paleopole locations coincide would be a step toward reconstructing their common history, especially if this is compatible with other geological indicators. The use of paleomagnetic data for continental reconstructions is especially successful when a sequence of coeval pole positions for each continental block is available.

Consider the paleomagnetic history of a tectonic plate that moves from a high latitude to a lower latitude and rotates as it does so (Fig. 6.3). Suppose that rocks form on the plate at three successive times, acquiring magnetizations with declinations D_1, D_2, and D_3, respectively. From the associated inclinations of the magnetizations I_1, I_2, and I_3 the corresponding polar distances p_1, p_2, and p_3 can be computed using Equation (6.1). Analysis of the paleomagnetic data yields the successive positions AP_1, AP_2, and AP_3, respectively, of the virtual magnetic pole (VGP). Viewed from a sampling location on the mobile plate, the paleomagnetic pole appears to have moved along a curved path during the time covered by the ages of the rocks, although in fact it is the plate that has moved. In this way, APW paths can be calculated for adjacent tectonic plates and used to reconstruct their earlier positions relative to each other.

APW paths were first constructed in the 1950s by geologists studying rocks on the North American and European continents. Confusingly, each continent was found to produce a different APW path. The Earth only has a single magnetic field; thus, investigators interpreted the paleomagnetic data as evidence for separate motions of the

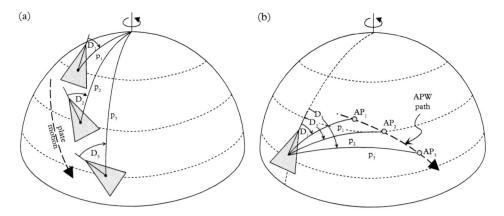

Fig. 6.3 *The formation of an apparent polar wander path (APW) for a moving tectonic plate. (a) As the plate moves southward, rocks become magnetized at different times with inclinations corresponding to the distance* p *from the pole (rotation axis) and declinations* D *determined by the degree of rotation at each position. (b) At a later time, the apparent pole position is calculated for each stage of the motion and defines the apparent polar wander path relative to the plate.*

individual continents, relative to the rotation axis and to each other, during their geological history. The slow motions were called continental drift, and the concept attracted adherents and opponents. The idea of solid continents apparently moving through solid oceanic crust defied logic, and how this could happen was a mystery that persisted for decades. The hypothesis of seafloor spreading developed in the early 1960s (Vine & Matthews, 1963) provided the mechanism. The "continental drift" was due to the relative motions of tectonic plates on which the continents are situated. Paleomagnetic investigations have documented plate motions by matching segments of APW paths for variously sized continental blocks.

The paleomagnetic method of reconstructing former plate motions is illustrated by the comparison of APW paths for Europe and North America (Fig. 6.4) for the long time interval between the Late Carboniferous period (450 Myr ago) and the Early Jurassic (170 Myr ago). A clockwise rotation of the European path by 38° about a pole—referred to as the Euler pole of rotation—located close to the present geographic pole brings the separate APW paths together in a very good fit. The APW paths match especially well for the hundred million years between the Late Carboniferous (310 Ma) and the Late Triassic (210 Ma). During this time, the European and North American continents formed a common tectonic plate, known as *Laurasia*. The rotated APW paths diverged in the Early Jurassic when the Atlantic Ocean began to open, moving the continents apart and separating the APW paths.

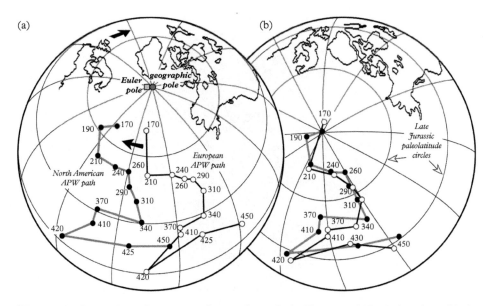

Fig. 6.4 *(a) Comparison of apparent polar wander paths for Europe and North America, with the continents in their present positions relative to each other. (b) A clockwise rotation of the European plate by 38° closes the Atlantic Ocean and causes the two APW paths to overlap. (Data source: R. Van der Voo, Phanerozoic paleomagnetic poles from Europe and North America and comparisons with continental reconstructions,* Reviews of Geophysics. *28, 167–206, 1990. With permission from John Wiley)*

The technique of analyzing and matching segments of APW paths for different continents has helped geologists to reconstruct in remarkable detail the relative locations of continents in the distant geological past, in particular since the Late Precambrian and Early Paleozoic. The interactions between continental blocks cause episodes of metamorphism and orogenesis that produce geological events in stratigraphy, paleontology, and structural geology. The integration of paleomagnetic data with these other disciplines has made it possible to reconstruct how continental blocks came together to form supercontinents and to understand their subsequent breakup. The following "thumbnail sketch" explains broadly how earlier supercontinents were formed.

In the late Proterozoic (about 800 Myr ago), the continental blocks that existed at that time started to merge to form larger landmasses separated by new oceans. From 800 Ma to 530 Ma collisions between India, Africa, South America, Australia, and Antarctica resulted in the formation of a single supercontinent called Gondwana in the southern hemisphere. Meanwhile, landmasses that are now predominantly in the northern hemisphere merged into a continental block called Baltica, composed largely of Scandinavia and Central Europe. Around 420 Myr ago, Baltica coalesced with Laurentia—the cratonic block that forms the core of North America.—to form the supercontinent of Laurussia (also called Euramerica). By the Late Permian to Early Triassic (at about 300 Ma), the assemblage of these units as a single supercontinent called Pangea was essentially complete (Fig. 6.5). Pangea existed as an entity for the following 100 Myr.

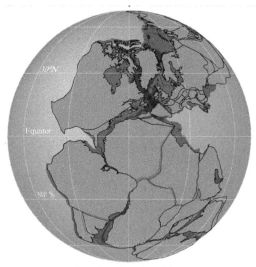

Late Permian (260 Ma)

Fig. 6.5 *The Permo-Triassic configuration of Pangea at 260 Ma in the Late Permian. (After Fig. 23 in M. Domeier, R. Van der Voo, and T. H. Torsvik, 2012. Paleomagnetism and Pangea: The road to reconciliation.* Tectonophysics, *514–517, 14–43. With permission from Elsevier)*

It was bordered by the Tethys Ocean to the east and the Panthalassa Ocean to the west. In the Late Triassic to Early Jurassic (about 200 Myr ago), Pangea began to break up, when the modern episode of seafloor spreading caused the Central Atlantic Ocean to open between Africa and North America. Subsequently, about 83 Myr ago, the North Atlantic opened between North America and Europe.

For the time interval since the Early Jurassic, plate motions can also be reconstructed with marine records of paleomagnetic reversals. The polarity sequence is recorded in the oceanic crust that has formed since the onset of the present phase of seafloor spreading in the Early Jurassic (Fig. 6.8 and Section 6.6). However, for eras older than the present oceans, the matching of paleomagnetic APW paths is the only quantitative method by which the detailed paleogeography of earlier supercontinents can be reconstructed.

Inherent in the derivation of an APW path are the assumptions (1) that the geocentric axial dipole hypothesis is valid and (2) that the rocks can be dated sufficiently accurately to connect their VGP locations in chronological order. A geocentric axial dipole field is a good model for paleomagnetic data of Cenozoic and Late Mesozoic ages, when plate tectonic motions can be taken into account. It becomes increasingly difficult to verify that the GAD hypothesis was valid in earlier geological periods. Also, the older the rocks are, the more difficult it is to order the ages of their VGP by radiometric dating.

6.5 Geomagnetic Polarity Reversals

In 1906, a French geophysicist, Bernard Brunhes, reported that the magnetizations of lavas from the Massif Central in central France were magnetized in almost the opposite direction to the present geomagnetic field, which was defined to have a *normal* polarity. He inferred that the geomagnetic field must have had the opposite, *reverse* polarity at the time the lavas formed. This conclusion was supported in 1929 by the work of Motonori Matuyama, a Japanese geophysicist, who noted a correlation between the magnetic polarity of basaltic lavas of Early Pleistocene age (~2 Ma) and their stratigraphic position. Matuyama arrived at the same conclusion that Brunhes had come to earlier, that the magnetic field had reversed polarity in the past 2 million years.

The study of geomagnetic polarity received little further attention until after the Second World War. The postwar development of radiometric dating and improvements in paleomagnetic equipment led to further interest in the topic. Detailed investigations of the history of reversals in sequences of lava flows that could be dated radiometrically were carried out by different research groups in the 1950s and 1960s. Gradually, the history of recent magnetic field reversals evolved and was extended back in time as dated lava sequences were studied in different geographic locations. The earliest of these field reversal studies were made before seafloor spreading was discovered. The paleomagnetists developed a nomenclature for describing and identifying intervals of constant normal and reverse polarity (Fig. 6.6). The longer "epochs" of constant polarity were named after pioneering scientists in geomagnetism (e.g., Brunhes, Matuyama), and short oppositely magnetized "events" within the "epoch" were named after the locations where they were first identified (e.g., the Olduvai gorge in Africa).

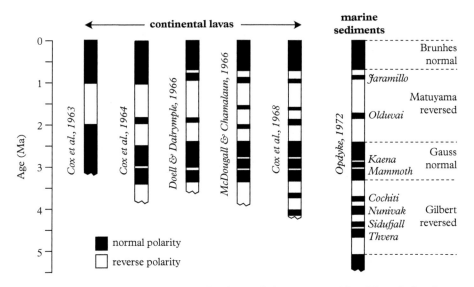

Fig. 6.6 *Evolution of the magnetic polarity time scale for the past 5 Myr. The polarity chrons established in continental lavas were dated radiometrically. The polarity history was confirmed and extended by correlation with the reversal sequence found in deep-sea sediments. (Data sources: Cox, Doell, & Dalrymple, 1963, 1964, 1968; Doell & Dalrymple, 1966; McDougall & Chamalaun, 1966; Opdyke, 1972)*

The sequence of polarity chrons in the past 5 Myr was corroborated in the late 1960s and early 1970s by independent analyses of magnetic polarity in marine sediment cores. The sediments were taken from deep ocean basins and sampled by using gravity to drive a long thin pipe, with a heavy mass on top of it, into the soft sediment. The magnetic records in the sediments provided an independent confirmation of geomagnetic polarity reversals because the remanent magnetizations of lavas and sediments are acquired by entirely different processes. The lavas are magnetized thermally, acquiring a TRM on cooling from high temperature, but the sediments acquire their remanent magnetization during deposition at ambient temperature. Yet the two media carried identical records of the changing polarity of the geomagnetic field. This countered the criticism of skeptics that a polarity reversal could simply be a mineralogical feature in the rock or sediment.

6.6   Magnetic Polarity Stratigraphy

Stratigraphy is the study of changes in the properties of rocks from one layer to another in a layered sequence. It is an important field of geology, in particular for understanding the history of sedimentary rocks. Magnetic stratigraphy (often shortened to *magnetostratigraphy*) is a variant of stratigraphy that makes use of changes in the magnetic properties of layered rocks and sediments. The most common form is magnetic polarity stratigraphy.

The direction of the remanent magnetization is measured, and from it the polarity of the magnetic field is obtained for the time when the original sediment was deposited. The field maintains a constant polarity for very long times, lasting from tens of thousands to tens of millions of years, whereas the transition from one polarity to the other only takes an estimated 5,000–7,000 years. The transition marks the sharp edges of blocks of constant polarity, as in Fig. 6.6. The method has been applied successfully to sedimentary rocks on land and to cores of sediment drilled from the ocean floor in the Deep Sea Drilling Project (DSDP), for example. The magnetic stratigraphies measured in land sections and oceanic drill cores of the same age agree well and complement each other. They extend the history of geomagnetic polarity from young, unconsolidated deep-sea sediments far back in geological time.

The ideal medium for magnetic stratigraphy is a sediment or sedimentary rock that was deposited in tranquil conditions. Fine-grained pelagic sediments that were slowly deposited in open ocean basins far from land meet the requirements. They usually contain very fine-grained magnetite as their primary magnetic mineral, which acquires a postdepositional magnetization close to the time of deposition. Hematite can form long after a sediment is deposited; if it forms much later, it may carry a secondary magnetization that overprints the direction of a primary pDRM. However, if the hematite magnetization is acquired close to deposition, this is not the case. The pelagic limestones in the Italian Apennines contain both magnetite and hematite, but both minerals were magnetized at about the same time, closely after deposition in the Cretaceous period (66–145 Myr ago).

Increasing pressure during burial of the sediment squeezes water out of the pore spaces, the grains become cemented during lithification, and the processes of diagenesis and compaction can change the original mineralogy of the rock, including the ferromagnetic fraction. As a result, the magnetization of rocks that were deposited on the ocean floor but now are exposed on land may have been altered by postdepositional physical and chemical changes or by later tectonic displacements. The laboratory and statistical methods used routinely in paleomagnetic investigations are used to "clean" any secondary components from the remanent magnetization and to define its original direction.

The magnetic polarity sequence derived from detailed, closely spaced sampling of the limestone in the Bottaccione gorge at Gubbio, Italy, illustrates the use of magnetostratigraphy to correlate and date the oceanic lineations produced by seafloor spreading (Fig. 6.7a). The paleomagnetic data from the top 150 m of the stratigraphic section in the gorge give a clear magnetic stratigraphy. The latitude of the virtual geomagnetic pole, calculated for each magnetically cleaned sample, is plotted against the stratigraphic position of the sample. Ideally, the VGP should vary between 90°N (present, normal field) and 90°S (reverse field). The observed directions have some scatter due to the difficulty of measuring the weakly magnetic limestones, but they show clearly the changes in polarity of the magnetic field during the deposition of the original sediment. Paleontological dating of the shells of tiny marine organisms (planktonic foraminifera and nannoplankton) that inhabited the seas at that time showed that the section was deposited in the Late Cretaceous epoch. Paleontology provides a biostratigraphic framework for the section and locates the stratigraphic positions of key stage boundaries.

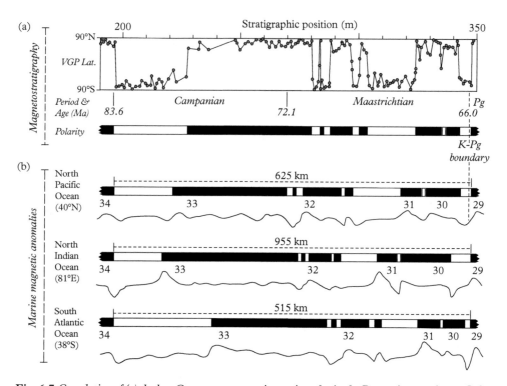

Fig. 6.7 *Correlation of (a) the late Cretaceous magnetic stratigraphy in the Bottaccione section at Gubbio, Italy, with (b) the polarity sequences interpreted from marine magnetic profiles in different oceanic basins. (Redrawn after: W. Lowrie and W. Alvarez, Late Cretaceous geomagnetic polarity sequence: detailed rock and paleomagnetic studies of the Scaglia Rossa limestone at Gubbio, Italy.* Geophysical Journal of the Royal Astronomical Society, *51, 561–581, 1977)*

The magnetostratigraphy at Gubbio correlates well with the reversal sequences derived from four marine magnetic profiles measured in different ocean basins on marine surveys that were hundreds of kilometers in length (Fig. 6.7b). The marine anomalies have different signatures because they were measured on magnetic surveys across ridges with different orientations to the magnetic field, but all result from the same polarity sequence. The reversal history recorded in the limestone section is clearly the same as that recorded in the ocean crust by seafloor spreading. The important biostratigraphic boundary between the Cretaceous and the Paleogene periods, which marks the mass extinctions caused by a huge asteroid impact, correlates with a time when the geomagnetic field had reverse polarity, in the uppermost part of polarity chron C29r.

The absolute ages of many paleontological stage boundaries are known from correlations to radiometric age dates that were obtained in igneous rocks and minerals. Many of the dates for stage boundaries have been refined by *cyclostratigraphy*, which can date sedimentary rocks by making use of cyclical changes in stratigraphic properties produced by fluctuations in the Earth's rotation and its orbit (see Chapter 7). The ages of the magnetic reversals between the dated stage boundaries can then be calculated by interpolation. In this way, a reversal sequence can be converted to a geomagnetic polarity time scale (GPTS) in which each reversal is accorded an absolute age. In practice, this is a complicated undertaking. A representative polarity sequence must first be determined for the oceanic crust. The magnetic anomalies on marine magnetic survey lines show the same succession of polarity chrons over similarly aged crust, although there are small differences in the relative lengths of a particular chron from one profile to another (Fig. 6.7b). The discrepancies are due to small irregularities in spreading rate at the ridges. The construction of a GPTS requires unifying such differences for a large number of profiles and optimizing the ages of the magnetic reversals. This was first achieved for the reversal sequence that extends between the new anomaly being formed at a modern ridge axis to the young edge of Cretaceous anomaly 34, which has an age of 83 Ma (Cande & Kent, 1992). Gradually, research on marine anomalies and magnetic stratigraphy has extended the known history of reversals back in time to the middle Jurassic, about 155 Myr ago.

The interpretation of magnetic lineations in terms of the age of oceanic crust makes it possible to reconstruct the history of how an ocean basin developed. All locations along a lineated anomaly were formed at the same time, so each lineation is an *isochron*. Magnetic lineations of the same age, which now lie on adjacent lithospheric plates on opposite sides of a ridge, were formed together at the ridge. By fitting successive pairs of equally aged lineations to each other, the opening of the ocean basin can be reconstructed. A classic example of this is the history of the opening of the Atlantic Ocean (Fig. 6.8). The reconstruction was derived by the statistical fitting of equally old magnetic lineations across the Mid-Atlantic Ridge. The relative positions of the plate boundaries at a number of stages in the opening of the Atlantic Ocean show that it was initiated by Africa separating from North America about 180 Myr ago in the Early Jurassic. An Oxfordian age of 155 Ma for the young margin of the Jurassic Quiet Zone in the western North Atlantic was estimated from paleontological examination of the bottom sediments overlying igneous oceanic basement recovered by deep drilling at DSDP Site 105 in the western North Atlantic.

The history of geomagnetic polarity before the present episode of seafloor spreading began must be derived without the continuous polarity record provided by oceanic magnetic anomalies. Magnetic stratigraphy in sedimentary rocks exposed on land can provide the polarity history of the magnetic field in the Early Mesozoic and older eras. Progress in finding the absolute ages of sequences of sedimentary rocks has enabled accurate dating of the polarity record.

6.7 Geomagnetic Polarity in the Early Mesozoic and Paleozoic

Small variations in the Earth's astronomical parameters modulate the seasonal amount of sunlight that falls on any place on Earth (i.e., its *insolation*), causing long-term cyclical changes in the climate. The cycles that drive the changes arise because (1) the planet's

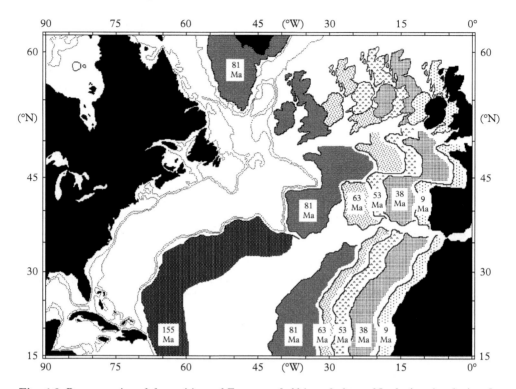

Fig. 6.8 *Reconstruction of the positions of Europe and Africa relative to North America during the opening of the Central and North Atlantic Oceans. Numbers on the European and African plates are their estimated ages at the indicated positions. (After: W. C. Pitman III and M. Talwani, Sea-floor spreading in the North Atlantic.* Geological Society of America Bulletin, *83, 619–646, 1972. With permission)*

shape is not a perfect sphere, (2) the rotation axis is tilted to the plane of its orbit, and (3) the shape of the orbit is slightly elliptical. As a result, the orientation of Earth's rotation axis wobbles slightly about the pole to the ecliptic (called axial precession) and nods up and down while it does so (called axial nutation). Meanwhile, the shape of the orbit (called its eccentricity) changes from nearly circular to about 6% elongate, and its long axis moves slowly around the ecliptic (called apsidal precession).

The changes are cyclical and are known collectively as the *Milankovitch cycles*. They affect the insolation and climate with periods of 21 ky, 41 ky, 100 ky, and 405 ky. The long-period 405 ky cycle in orbital eccentricity is controlled by the enormous mass of Jupiter and the nearness of Venus and is very stable. It is referred to as the "metronome" for the astronomical dating of ancient sedimentary sequences.

By modulating the insolation, the rotational and orbital cycles modify the global climate. The climatic effects modulate temperatures and precipitation, which affect sedimentation rates and the subsequent thicknesses of sedimentary layers. This has given rise to the stratigraphic technique of *cyclostratigraphy*, which has grown in importance in the past 30 years as a way to determine the absolute ages of sedimentary sequences. By counting and measuring the layer thicknesses, the absolute ages of sedimentary sequences can be determined. The different periods of the Milankovitch cycles over-lap and interfere with each other, thus modulating sedimentation and layer thickness. Frequency analysis can resolve the interaction and relate the occurrence of periodici-ties in sedimentation to the basic cycles. The 405 ky "metronome" cycle is particularly useful in very old rocks because of its long-term stability.

Cyclostratigraphy has become an important tool for dating layered rocks and has greatly improved the calibration of the geological time scale. It has permitted the history of geomagnetic polarity to be dated for geological epochs that are older than the present ocean floor, in particular by extending the Cenozoic and Late Mesozoic record into the Late Triassic.

The history of geomagnetic polarity reversals during the Late Triassic and Early Jurassic has been derived in a landmark study of nonmarine sedimentary rocks (Olsen & Kent, 1999). The redbeds were deposited as lacustrine sediments in the Newark Basin in the eastern United States. By matching polarity zones in seven overlapping drill cores in the redbeds (Fig. 6.9) the investigators were able to piece together a well-defined his-tory of geomagnetic polarity in the Late Triassic epoch. An absolute age of 202 Ma for the Triassic/Jurassic boundary provides an anchor point at the top of the magnetic stratigraphy.

The overlapping drill cores in the Newark redbeds were dated by cyclostratigraphy based on rhythmic fluctuations in lithology. The most important ones are driven by the 405 ky eccentricity cycle and are known as McLaughlin cycles. In conjunction with the absolute age of the anchor point, it was possible by this means to extend the known his-tory of geomagnetic reversals well into the Late Triassic epoch, covering the age interval 202–233 Ma. Parts of the Newark record have since been confirmed independently in sections from Greenland, Turkey, and the Italian Alps.

The history of geomagnetic field polarity before the Late Triassic is less well known. The Newark polarity record is the oldest with a distinct magnetic stratigraphy that is also dated accurately. Geomagnetic polarity reversals in older rocks are less well documented. There are few continuous polarity records, but some broad features of older polarity history have been established.

The Late Permian was characterized by an interval of alternating normal and reverse polarity called the Illawarra reversals. It is uncertain when in the Triassic this mixed polarity interval ended. The age of onset of the reversals is estimated to be around 269–265 Ma, although there is some disagreement over the exact timing. Different attempts to establish a definitive magnetic stratigraphy for the Illawarra reversals have yielded inconsistent polarity patterns.

Paleomagnetic results indicate that during the Early Permian and Late Carboniferous (Pennsylvanian)—that is, between about 267 Ma and 318 Ma—the magnetic field had a constant reverse polarity for an exceptionally long time. Apart from 3–4 possible short (unconfirmed) normal polarity chrons, it lasted about 50 My, which is much longer than the Cretaceous Normal Polarity Superchron. Named after the area at Kiama in Australia,

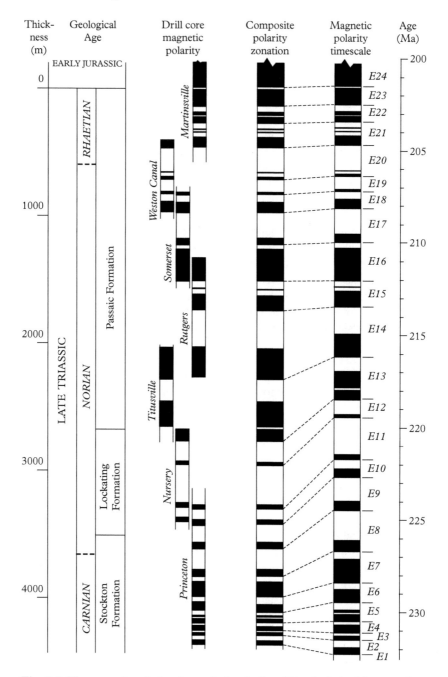

Fig. 6.9 *The magnetic polarity time scale for the Late Triassic derived from overlapping drill cores in the Newark basin, dated by cyclostratigraphy.* (Data source: P. E. Olsen and D. V. Kent, *Long-period Milankovitch cycles from the Late Triassic and Early Jurassic of eastern North America and their implications for the calibration of the Early Mesozoic time scale and the long-term behavior of the planets.* Philosophical Transactions of the Royal Society London A, *357, 1761–1786, 1999*)

where it was first investigated, the interval of reverse polarity is known formally as the Kiaman Reverse Polarity Superchron.

A large number of magnetostratigraphic sections have been measured in Early Ordovician and Late Cambrian rocks (ages 470–510 Ma) from the Siberian platform. A synthesis of the results produced a magnetic polarity time scale for this interval of the Early Paleozoic. The polarity sequence shows well-defined normal and reverse polarity zones, as well as some zones characterized by only one or two samples. The investigations suggest the occurrence of a possible superchron with reverse polarity, estimated to have lasted ~20 Myr, in the Early to Middle Ordovician (around 470 Myr ago). However, the polarity sequences correlated only weakly between profiles from different geographic regions. The Siberian results emphasize how difficult it is to establish a coherent history of geomagnetic behavior in the Earth's very distant past.

6.8 The Geomagnetic Field in the Precambrian

The Precambrian encompasses approximately 88% of the age of the Earth. It covers the interval of time from the base of the Cambrian period at 541 Ma back to the planet's formation at 4540 Ma. Its history is subdivided into three unequal time units called eons. The oldest is the Hadean, a lifeless eon about which little is known, as it left few traces. It represents the time between the origin of the hot, molten planet and the beginning of the Archean eon. No rocks survive from this time, but zircon grains of Hadean age have been incorporated in some younger metamorphosed rocks. Any primordial atmosphere was probably stripped from the Earth by the fierce solar wind before the geomagnetic field could develop as a protective shield. The Archean eon lasted from 4,000 Myr ago until 2,500 Myr ago; it is characterized by the earliest rocks in which primitive life forms have been found. Continental masses may have existed during the Archean, but they have been consumed by subsequent plate motions, leaving only remnants that form the ancient cratons on modern continents. The Proterozoic eon began 2,500 Myr ago and is the longest eon in the geological time scale. Early in this eon, photosynthesis by bacteria in the oceans released oxygen, which was absorbed by oxidizing active elements such as iron, forming banded layers of iron oxides that are hundreds of meters thick and extend for hundreds of kilometers. In the late Proterozoic, this activity ceased, and the oxygen accumulated to form the tertiary atmosphere that continued until the present day. Modern plate tectonic motions are thought to have started during the Proterozoic and to have produced early supercontinents or assemblages of continental blocks. The Proterozoic lasted almost 2,000 Myr until the start of the Phanerozoic eon, which is composed of the Paleozoic, Mesozoic, and Cenozoic eras (Fig. 6.10).

The Geocentric Axial Dipole hypothesis has not been established with the same degree of confidence in the Precambrian as it has for the Cenozoic and Mesozoic eras. It is not even certain that the Precambrian magnetic field was generated by the same type of geodynamo as the present field. For example, it is not known when the Earth's solid inner core began to nucleate. The field lines generated by a dynamo in the liquid

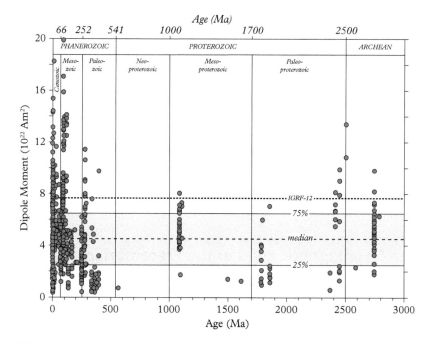

Fig. 6.10 *Paleointensity results for the Precambrian and the Phanerozoic eons. The data are selected from the PINT paleointensity database (v.8.0.0, 2022) and pass at least 3 of the 10 criteria for reliability. Additionally, they must have a well-dated age, lack signs of alteration, and have magnetizations that are not carried by multidomain-sized magnetite.*

outer core field diffuse into the conducting solid inner core, which is believed to have a stabilizing effect on the dipole field. Prior to the formation of an inner core the geomagnetic field may have differed from the configuration of the historic field. The early field may have had stronger nondipolar components at the Earth's surface than would be produced by a geodynamo stabilized by the inner core. If the nondipolar fields were large at the Earth's surface, they could invalidate the GAD hypothesis, which would have serious implications. Properties calculated for a dipolar field, such as its virtual dipole moment (VDM) and the location of the virtual geomagnetic pole (VGP), would lose their meaning.

Several paleomagnetic studies in Precambrian rocks, however, have obtained results indicating that the field was indeed predominantly dipolar. The magnetic stratigraphy in 1,750 Myr sedimentary rocks from the Northwest Territories in Canada (Bingham & Evans, 1976) displayed a clear polarity zonation, with antipodal normal and reverse directions. The scatter in the directions was large, but their stratigraphic organization into oppositely magnetized polarity zones suggests that the field at that time in the Precambrian was probably dominantly dipolar and undergoing reversals in the same way as it has done throughout the Phanerozoic.

Additional confirmation of the GAD hypothesis has been obtained from an analysis of antipodal directions in the paleomagnetic database (Veikkolainen, Pesonen, & Korhonen, 2014). The directions of normal (N) and reversely (R) magnetized samples were compared from 214 studies, in which both polarities were measured at the same site. For a GAD field, the N and R directions from a common site are antiparallel to each other; that is, they are 180° apart. The analysis of Precambrian data in the database showed that the directions of the N and R polarities were well fitted by a dipole field that was accompanied by quadrupole and octupole components measuring 4% and 5% of the GAD, respectively. The nondipole components are similar in size to those interpreted from the magnetization inclinations of modern deep-sea sediments.

Paleomagnetism is able to compute APW paths and reconstruct supercontinents in the Phanerozoic eon that occupies the 541 Myr since the end of the Precambrian because the geomagnetic field was dominated by its axial dipole component during this time (Section 6.4). In contrast to the Paleozoic, where paleomagnetic and geological data agree that continental blocks came together to form the supercontinent Pangea (Fig. 6.5), the paleogeography of the Earth's surface is virtually unknown for most of the Precambrian. Fragmentary knowledge begins in the late Mesoproterozoic about 1,200–1,000 Myr ago, when several continental blocks came together and assembled a supercontinent called Rodinia. Paleomagnetic data of that age from different cratons has contributed to our understanding of late Precambrian paleogeography, although there is not unanimous agreement. Different paleogeographic reconstructions show different configurations of the constituent cratons in the supercontinent. Geological constraints and paleomagnetic data agree that Rodinia began to break up about 800 Myr ago, with final fragmentation about 600 Myr ago. The breakup was accompanied by the development of the Iapetus Ocean, which separated continental blocks called Baltica, Laurentia. and Amazonia. Iapetus was the predecessor of the modern Atlantic Ocean.

The paleointensity of the Precambrian field provides an alternative source of information about the internal dynamics of the early Earth. For example, the strength of the field could be affected by changes in the vigor of core convection, the thermal gradient across the liquid outer core, and the existence of an inner core. In order to compare paleointensities at different locations, it is necessary to convert them to the strength of the corresponding geocentric axial dipole—its virtual dipole moment (VDM), which assumes the GAD hypothesis. Several thousand paleointensity analyses on rocks with ages from 50 ka to 2744 Ma have been assembled in a unified database. The reliability of each entry in the database is classified with 10 criteria, which cover a range of tests. These include, for example, how well the age is known, whether alteration of the sample is known or suspected, and whether the grain size of the magnetic minerals is single domain or multidomain. A multidomain magnetization is unstable and would deliver a false paleointensity result. The paleointensity of the field is difficult to determine, and there are many places where errors can arise. The quality of paleointensity studies has steadily improved, but many entries in the paleointensity (PINT) database are of dubious value. The reliability criteria help select the best data.

A paleointensity value is judged as acceptable if it passes at least 3 of the 10 reliability criteria. Applying a further restriction, that the data must pass the criteria for well-dated age, lack of alteration, and absence of multidomain grain size, reduces the depleted database to about 900 values. The oldest data in the selection have ages of about 2750 Ma, but most of the data are Phanerozoic in age (younger than 540 Ma). Paleointensity data from the Precambrian are very sparse and cover a wide range of values (Fig. 6.10), which lie slightly below but are comparable to the Phanerozoic results. The median and quartile values of the dataset are not as strong as the recent field intensity, represented by the axial dipole component of the International Geomagnetic Reference Field IGRF-12.

The most noticeable features of the Precambrian paleointensities are the sparsity of reliable data compared to younger ages, the very large gaps in time between the dates of accepted analyses, and the large scatter in ages at each date. Some of the oldest data were dated by the Pb-Pb radiometric method and are very precise. However, a relative error of as little as 0.1% translates to an absolute error of ±2 Myr for a 2,000 Myr age. Thus, each of the narrow vertical distributions of ages may represent a number of different ages within such a time window. The broad range of intensities at each age may be real and indicative of large variations in strength of the paleomagnetic field, similar to those evident in the Phanerozoic data, for example in the past million years. However, at present, the quantity and quality of paleointensity data from the Precambrian are insufficient to establish when the solid inner core nucleated.

7

The Effects of Solar Activity on the Geomagnetic Field

Introduction

On September 1, 1859, a huge eruption occurred in the Sun's corona, the outermost layer of its atmosphere. It ejected an enormous mass of electrically charged particles in the direction of the Earth, which it reached less than 18 hours later. The coronal mass ejection (CME) was preceded by a short flare of white light that was observed and described by a British astronomer, Richard Carrington. The exceptional CME became known as the Carrington event. It caused massive disturbances of the Earth's magnetic field, creating the largest magnetic storm on record. It induced currents in the cables of the American telegraph network that damaged the system and put it out of operation. The infusion of charged particles into the Earth's atmosphere caused auroral displays around the world that were especially bright in high latitudes and visible even in semitropical latitudes. If a CME like the Carrington event occurred today, it would cause enormous damage.

The Sun constantly radiates electrically charged particles. They form a plasma called the *solar wind*, which carries the solar magnetic field far out into space. The radiation—in particular the extreme amount of energy associated with a CME—presents a major hazard to human life outside the Earth. In August 1972, a CME impacted the Earth and Moon. It arrived at the Moon between the Apollo 16 and 17 lunar landing missions. The timing was fortunate: the Moon has neither an atmosphere nor a global magnetic field to protect it, so if the CME had happened when astronauts were on the lunar surface, they would have been exposed to dangerous, possibly lethal, levels of radiation.

The solar wind and coronal mass ejections have a strong influence on the Earth's magnetic field in the near-space environment around the Earth. Without the protection of the geomagnetic field, the plasma radiation would be lethal for life on Earth. As it is, the solar emissions, described collectively by the term *space weather*, cause many problems for the infrastructure our civilization depends on. In order to anticipate extreme events, we need to understand the processes in the Sun that lead to the emission of electromagnetic and particle radiation.

The Earth's Magnetic Field. William Lowrie, Oxford University Press. © William Lowrie (2023).
DOI: 10.1093/oso/9780192862679.003.0007

7.1   The Internal Structure of the Sun

The Sun is a middle-aged star that is about halfway through its life cycle. Its age is believed to be about 4.6 billion years old, based on independent estimates from radiometric dating of meteorites and from astronomic modeling. It is composed largely of hydrogen and helium, with minor amounts of other elements. Its size is gigantic in comparison with any of the planets that orbit it. The visible solar disc has a diameter of 1.39 million km, which makes it 109 times larger than the Earth and 10 times larger than Jupiter, the largest planet in the solar system. The Sun's mass is 333,000 times that of the Earth and 1,050 times that of Jupiter; it is equivalent to more than 99.8% of the cumulative mass of all the planets, satellites, asteroids, comets and other objects in the solar system. The average distance of the Earth from the Sun is called an astronomical unit (AU); it measures approximately 150 million km. However, due to the motion of the Earth around its elliptical orbit, the distance from the Sun changes throughout a year. A more exact definition of the unit, based on celestial mechanics and involving compensation for relativistic effects, is now used. As a result, the AU is now defined to be exactly 149,597,870.7 km.

The Sun is gaseous and thus does not have a firm surface. Instead, its surface is equated with the *photosphere*, which is a thin outer shell about 100 km thick that forms the visible disc we see from the Earth. It is not possible to look directly into the interior of the Sun beneath the photosphere. However, the surface of the photosphere is seen in close-up to be covered by millions of small light areas bounded by narrow dark zones. This feature, called *granulation*, gives the photosphere a mottled appearance (Fig. 7.1) and provides clues to internal processes in the Sun. The light areas (granules) only appear small because of the huge area of the solar disk; in fact, they can be thousands of kilometers in diameter. They represent the tops of columns of hot gases that rise through the convective zone, spread out at the photosphere and cool, before sinking back into the Sun at the cooler, darker margins. The granules last only 5–10 minutes. However, just below the photosphere, at the top of the convective zone, there is a layer of large supergranules: these may be 30,000 km across and can last up to 24 hours. The hot columns rise at speeds of 2–3 km/s through the convective zone and cause the photosphere to move up and down where they reach the surface. The oscillations of the visible surface alter the wavelengths of light emitted, which changes the spectrum of their frequencies. The changes are observed by analysis of spectrograms obtained by telescope from the Earth and from dedicated spacecraft.

Little was known about the rotation of the Sun's interior until the development of *helioseismology*, which has led to a deeper understanding of the Sun's large-scale internal structure as well as its internal rotation. Helioseismology is based on the analysis of oscillations of the photosphere, which are caused by acoustic waves that bounce around in the solar interior. The gaseous Sun does not support shear waves, but compressional waves travel through the Sun in the same way as terrestrial acoustic waves, being reflected and refracted (i.e., changing direction) at major interfaces. A wave with a particular period may become trapped by multiple reflections in a suitable volume or cavity.

Fig. 7.1 *Image of a sunspot and surrounding granulation on the Sun's photosphere, observed with the Swedish 1-m Solar Telescope on La Palma, Canary Islands. The solar magnetic field is vertical at the dark center, or umbra, and more horizontal in the adjacent area. The superposed image of the Earth (diameter 12,800 km) gives an impression of the colossal size of features on the Sun.* (Image credit: *The Royal Swedish Academy of Sciences, V. M. J. Henriques, D. Kiselman, and NASA (Earth Image)*

A trapped acoustic wave reinforces itself by constructive interference at a given period, which causes it to resonate. By observing a given location on the solar surface for minutes at a time, a record of the oscillations is obtained from which the resonant modes are then derived. The surface oscillations are produced by the superposition of more than a million resonant modes. A supercomputer is needed to analyze them.

The internal structure is determined by combining analysis of the surface oscillations with theoretical modeling. A computer model of the interior of the Sun, referred to as the *Standard Solar Model*, has been developed that is consistent with its observed properties. The internal structure is assumed to consist of concentric spherical shells like that of an onion. With increasing radial distance from the Sun's center, these are successively the core, the radiative zone, and the convective zone (Fig. 7.2). A large number of parameters are included in the numerical model (e.g., mass fractions of hydrogen and

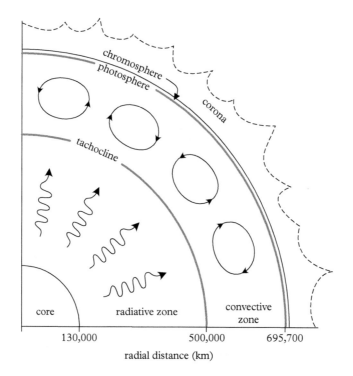

Fig. 7.2 *Sketch of the layered internal structure of the Sun. The photosphere, chromosphere, and corona form the Sun's atmosphere. The corona is much greater than depicted and extends for millions of kilometers into space.*

helium, density, pressure, temperature, energy generation per unit of mass). To construct the model, the parameters are altered slightly from each layer to the next, taking into account the issue of whether the layer is in the core, radiative zone, or convective zone. In this way, a model of the radial variations of physical parameters is built up.

7.2   Energy Transfer in the Sun

The energy of the Sun is produced in the central *core*, where nuclear fusion converts hydrogen to helium at a temperature of more than 15 million degrees. The temperature is so high that electrons can escape from their atoms, converting the atoms to positively charged *ions*. These coexist with the negatively charged free electrons in the form of a hot ionized gas, or *plasma*. In the fusion reaction, the mass of each helium nucleus is less than the total mass of the four hydrogen nuclei that fuse to form it. The difference in mass is emitted as electromagnetic energy, largely in the form of gamma radiation, which heats up the Sun through collisions with protons and electrons in the solar plasma.

The emissions of electromagnetic energy have been taking place since the solar system formed about 4.6 billion years ago. In that time, much of the Sun's original hydrogen has been converted into a central core of helium. The temperature and pressure within the core are not high enough to cause further fusion of the helium, which forms a slowly growing but stable central core.

Gamma rays from the solar nuclear reaction do not escape from the Sun. The thermonuclear core is surrounded by a *radiative zone*, in which the density of matter is very high. Although the gamma radiation travels outward from the core at the speed of light, the high density of the zone results in frequent collisions between the *photons* of gamma radiation and other solar particles, slowing down dramatically the transmission of the radiated energy. The electromagnetic radiation is absorbed within a short distance by these particles, which absorb and reemit it at a slightly longer wavelength. This causes the temperature to decrease from about 15 million degrees at the base of the zone to 1.5 million degrees at its top. As a result of the collisions between the photons and solar particles, the electromagnetic energy passes very slowly through the radiative zone.

The outer layer of the Sun is the *convective zone* in which energy is transferred dominantly by convection. Within the radiative zone, the temperature gradient is too low for convection to take place. In the convective zone, on the other hand, the plasma has cooled enough so that it is no longer sufficiently hot to transfer energy by radiation. Convection then becomes the prevailing mode of energy transfer in this outer zone of the Sun. The nature of convection in the Sun is turbulent and violent, and has the effect of cooling the star to a surface temperature of around 5,500°C, which is comparable to the temperature at the center of the Earth.

The transport of electromagnetic energy from the core to the Sun's surface is slow, taking millions of years, but when the energy reaches the photosphere, electromagnetic waves with a very broad spectrum of frequencies (Chapter 1.8) radiate out into space at the speed of light. Energy is also transported out of the Sun by electrically charged particles, mostly protons and electrons of the solar plasma, but at a slower rate. As a result of the wave-particle duality principle of atomic physics, this process is regarded as *particle radiation*. When it reaches the Earth, it has a larger effect on the geomagnetic field than the wave-like radiation.

Outside the photosphere is a spherical layer, about 3,000 km thick, called the *chromosphere*. Its density is very low, about 10,000 times less than that of the underlying photosphere, which is so much brighter that it makes the chromosphere invisible. Without special equipment it can only be seen during a solar eclipse, when it gives the Sun a reddish color.

The outermost layer of the Sun is a low-density plasma, called the *corona*, that extends into space for millions of kilometers. The corona is much hotter than the photosphere and chromosphere, and it typically has a temperature of several millions of degrees. The mechanism by which the corona is heated to such an extreme temperature is a target of ongoing research. In the corona, positively charged nuclei of hydrogen (protons) and helium (alpha particles) and negatively charged electrons are accelerated to speeds that exceed the escape velocity of the Sun. The plasma spreads out into space to form the *solar wind*, which is enriched spasmodically by increased emissions of plasma from the corona called *coronal mass ejections*. These violent events are often accompanied by a bright flash on the Sun's surface called a *solar flare*. The solar wind is a flow of electrically charged particles, and so it carries with it an *interplanetary magnetic field* (IMF), also referred to as the *heliospheric magnetic field*. The IMF interacts with and modifies the magnetic field around each planet or moon in the solar system that possesses an internally generated field.

7.3   Sunspots and the Solar Cycle

As long ago as the pre-Christian era, tiny dark patches on the face of the Sun were observed with the naked eye by Greek and Chinese astronomers. In 1613 *Galileo Galilei* studied them with his telescope and noted that they moved across the face of the Sun. His observation of sunspot motion is the earliest documentation that the Sun rotates.

The occurrence of a sunspot is a result of internal activity in the Sun. The motions of the plasma inside the Sun produce magnetic fields that have a strong effect on the internal processes. In places, the local magnetic fields are strong enough (on the order of 0.3 T) to prevent light from escaping from the Sun. This causes a dark region on the surface of the photosphere that is known as a sunspot (Fig. 7.1). In these regions the magnetic field inhibits thermal convection and lowers the local temperature to form a "cool" area with a temperature of around 3,500°C, which is about 2,000°C cooler than its surroundings. The dark regions are referred to as "spots" because they appear to be small features, but this is only in relation to the huge size of the Sun. In reality, the diameter of many sunspots is comparable to or greater than the diameter of the Earth.

Sunspots occur at places where the Sun's corona is linked by the magnetic field to the deep interior below the photosphere. The number of sunspots at any given time is determined by the current state of the Sun's internal activity. It rises and falls with a period of about 11 years, which is known as the *sunspot cycle* (Fig. 7.3). The changes are not symmetric: the number of sunspots increases from minimum to maximum more quickly than it decreases back to minimum. The sunspots occur in pairs, with opposite magnetic polarities, one in each of two bands of latitude within 40 degrees north and south of the solar equator. At the beginning of a sunspot cycle, a member of each pair first appears in the outer part of each band. As the cycle progresses, each band broadens and migrates toward the equator. When the sunspot locations are plotted as a function of time, they form a so-called butterfly diagram. Because a sunspot is the result of magnetic activity and is anchored to underlying electrical currents, the butterfly diagram is interpreted as evidence for flow of plasma in the Sun's convective zone toward the equator in both hemispheres.

Spectrographic analysis of light emitted from different locations on the surface of the Sun confirms the rotation of the Sun's surface inferred from the motion of sunspots. Viewed from above the north pole of the Earth, the Sun and the planets that orbit it rotate in the same counterclockwise sense. Astronomic observations of the Sun's rate of rotation make use of the Doppler effect, which is familiar as the change in frequency

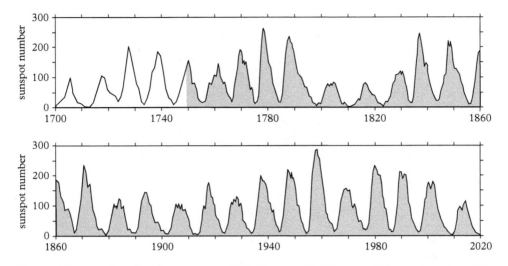

Fig. 7.3 *The annual number of sunspots since 1700 illustrates the 11-year cyclicity of solar activity. The data are annual means before 1749 and smoothed monthly means after 1749. (Data source: Sunspot Index and Long-term Solar Observations (SILSO), Royal Observatory of Belgium, Brussels)*

of a periodic signal when it is moving toward or away from an observer (a frequently quoted example is the changing pitch of a passing police siren). When the source is moving toward the observer, the effective wavelength of the signal is shortened and the frequency is raised. Conversely, the frequency is lowered when the source is moving away from the observer.

Spectrographic analysis of light emitted by hydrogen atoms at the edges of the solar disk—called its limbs—shows that the spectral lines are shifted slightly to higher frequencies at the approaching limb. At the receding limb, the spectral lines are shifted slightly to lower frequencies. From these slight changes in frequency the rotational velocity of the surface is calculated, and from that the rotational period is obtained.

Analysis of the Doppler effect on the spectral lines of hydrogen shows that the Sun's surface rotates with an average period of 28 Earth-days. However, the rotation rate is not uniform and varies with latitude (Fig. 7.4). It is fastest at the solar equator, where the sidereal period of rotation (i.e., the time required for a full rotation back to a fixed direction in space) is approximately 25 Earth-days; the corresponding synodic period (i.e., the time needed for a solar feature to return to the same position viewed from Earth) is about 27 Earth-days. The rate of rotation becomes slower with increasing latitude, and the sidereal period measures more than 30 Earth-days in high latitudes.

The latitudinal dependence of the rate of rotation has been shown by helioseismology to persist deep into the Sun's interior (Fig. 7.4). With increasing depth in the convective zone, the rotation rate remains faster at the equator than at the poles, but the difference decreases progressively. In the radiative zone, the differential rotational rate largely disappears, and the radiative zone appears to rotate as a solid body.

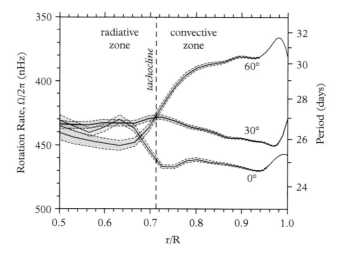

Fig. 7.4 *Variation of solar rotation with latitude and with depth in the Sun's interior, measured by the Michelson Doppler Imager (MDI) instrument aboard the Solar and Heliospheric Observatory (SOHO) spacecraft. The number on each curve is the latitude at which the rotation was measured. (Data source: Figure 13 in A. G. Kosovichev, J. Schou, et al., 1997. Structure and rotation of the solar interior: initial results from the MDI Medium-L program. Solar Physics, 170, 43–61)*

7.4   The Sun's Magnetic Field

The origin of the Sun's magnetic field is still incompletely understood. At times, the magnetic field has north and south magnetic poles; its magnetic equator is close to the ecliptic plane and slightly inclined to it. However, the external appearance of the Sun's magnetic field is not simply that of a dipole, but is usually very complex, with tangled magnetic field lines at all latitudes. The geometry of the field lines changes with time, reflecting systematic changes in the Sun's internal activity. As in the case of the geomagnetic field, internal rotation and motions of the electrically conducting plasma are important factors in generating the solar magnetic field. The appearance of the field changes markedly and regularly in the space of a few years and is correlated with the phases of the sunspot cycle.

At times when the solar activity is low, the number of sunspots is minimum, and the Sun's large-scale magnetic field approximates most closely that of a dipole (Fig. 7.5), with field lines entering the surface at one "pole" and leaving at the opposite one. At this stage in the solar cycle, the axis of the dipole is tilted slightly relative to the Sun's rotation axis. The tilt increases as the solar cycle progresses. As the number of sunspots increases, so does the complexity of the field. At times of highest activity, when the sunspot number is maximum, the magnetic field lines are very tangled, and there is stronger evidence for

2013 Sep 01

Fig. 7.5 *Model of the Sun's magnetic field at a time of low solar activity. The green field lines (outward directed) and violet field lines (inward directed) are "open": they connect with distant magnetic fields in the solar system. The white magnetic field lines are "closed"; that is, they leave and then return to the solar surface. (*Image credit: *Scientific Visualization Studio, NASA's Goddard Space Flight Center, https://svs.gsfc.nasa.gov/4124)*

a toroidal component. At solar maximum, the magnetic field is not well represented by a dipole. Subsequently, the field changes polarity and then becomes more dipolar in the opposite direction. A full polarity cycle—from normal to reverse and back—lasts 22 years, twice the period of the sunspot cycle.

The radiative zone and the convective zone are separated by a thin transitional layer known as the *tachocline*, which is thought to play an important role in generating the Sun's magnetic field. The steep velocity gradient in this layer results in a shearing motion in which overlying layers of the plasma move over lower layers, so that the plasma in the tachocline has a component of flow around the Sun. The magnetic field lines associated with this current are carried along by the plasma to form a toroid, or "belt," around the Sun. Estimates of the strength of the toroidal field indicate that it may measure about 10 T. For comparison, the superconducting magnet in the Large Hadron Collider at CERN, the European Centre for Nuclear Research at Geneva, achieves about 4 T.

Toroidal field lines may "leak" from their circumsolar girdle in the tachocline and thereby contribute to the Sun's external magnetic field. However, the internal structures of stars that are cooler than the Sun have no radiative zone or tachocline, yet these stars also have magnetic fields. So, it is likely that motions of the conducting plasma in the convective zone may be the major factor in generating the Sun's magnetic field. Although the origin of the solar magnetic field is still obscure, it appears to be generated by a dynamo mechanism like that in the Earth.

The Sun's magnetic field measures about 100,000 nT at its surface. By comparison, the magnetic field at Earth's surface measures about 30–60,000 nT. The solar magnetic field extends far into space and defines the outer limit of the solar system. The strength of a dipole field decreases according to the inverse cube of distance; thus, the solar magnetic field at the Earth's orbit should measure around 0.01 nT. However, the measured field at that distance is about 6 nT, hundreds of times stronger than expected. The difference is due to the solar wind, which is a flow of electrical charges and has its own magnetic field. It strengthens the solar magnetic field and carries it far into space beyond the Earth.

The region of space occupied by the Sun's magnetic field is called the *heliosphere*. At its outer boundary—the *heliopause*—the combined magnetic fields of the Sun and solar wind are balanced by the magnetic field of the interstellar medium, that is, the charged particles that occupy the vast space between the stars. The heliopause is at a distance of approximately 123 AU (18 billion km) from the Sun. This is far beyond the outermost planet in the solar system, the dwarf planet Pluto, which is around 40 AU from the Sun. The distance to the heliopause is not constant, due to variabilities in the velocity and the density of the solar wind and the interstellar medium. Effectively, it marks the outer limit of the solar system.

In 2012 the Voyager 1 spacecraft became the first human-made object to cross the heliopause and enter interstellar space, followed shortly thereafter by its companion spacecraft, Voyager 2. The spacecraft are traveling at around 17 km/sec and 15 km/sec, respectively, as they move ever deeper into interstellar space. They still send data back to Earth, but they are now so distant that the transit time of their signals, even traveling at the speed of light, takes around 17–20 hours.

7.5 The Solar Wind

The solar wind is a plasma composed mainly of electrons and protons, although about 5% consists of alpha particles and other ions. The density of the plasma is low; there are only on the order of five protons per cubic centimeter. However, the temperature of the particles is very high—more than 100,000 °C—so they have too much energy to combine with particles of opposite charge, and this enables them to coexist as a plasma. It is an electrically conducting fluid, whose motion and magnetic properties are subject to the rules of *magnetohydrodynamics* (Chapter 4.3). Electrical currents in the plasma create a magnetic field that is carried along with the solar wind as it flows through space. It combines with the Sun's magnetic field to form the IMF, so that it has a strength of about 6 nT at the Earth's orbit.

The charged particles in a plasma are coupled by the electrical conductivity, which makes it possible for magnetohydrodynamic waves to propagate in the plasma. One type of wave propagates within the plasma by pressure differences, like a sound wave in air, and is called the *magnetosonic* wave. Another wave propagates as oscillations of the magnetic field lines in the plasma. It is called an *Alfvén wave* in recognition of Hannes Alfvén, the Swedish astrophysicist who founded the discipline of magnetohydrodynamics.

Two types of solar wind are recognized, distinguished by their speed and internal temperature. The difference arises because plasma is emitted at variable speeds from different parts of the corona and at different phases of the solar cycle. The fast solar wind has a high temperature of ~800 million kelvins (K) and propagates at speeds of more than 700 km/s, while the slow solar wind is cooler, with a temperature of ~100 million K, and travels at around 300 km/s (Fig. 7.6). As it spreads out from the Sun, the fast wind overtakes the slow wind and transfers momentum and energy to it by compressing it and increasing the plasma density and temperature. The interaction of the fast and slow winds means that their speed at the Earth's orbit is very variable. Its median value is around 440 km/s. This is faster than the speeds of magnetosonic waves and Alfvén waves in the plasma, so the solar wind is both supersonic and super-Alfvénic. It takes about 2–6 days for the solar wind to travel from the Sun to the Earth; for comparison, visible light takes approximately 500 seconds.

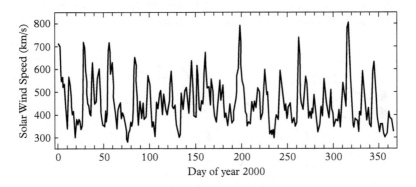

Fig. 7.6 *The speed of the solar wind at 1 AU, the distance of Earth's orbit from the Sun. (Redrawn after: Fig. 4a in J. T. Gosling, R. M. Skoug, and D. J. McComas, Low-energy solar electron bursts and solar wind stream structure at 1 AU,* Journal of Geophysical Research, *109, A04104, 2004. https://doi.org/10.1029/2003JA010309)*

7.6 The Interplanetary Magnetic Field

The solar wind dramatically modifies the appearance of the Sun's magnetic field in interplanetary space. When emitted from the corona the plasma has a radial velocity, and it also has a tangential velocity due to solar rotation. This leads to the formation of spiral-shaped flow lines of the escaping plasma. The spiral-shaped flow pattern is called a Parker spiral, in recognition of Eugene Parker, an American astrophysicist, who predicted the pattern in 1958. It forms in the same way as an Archimedean spiral, as follows.

Consider the emission of plasma from a point A on the coronal surface (Fig. 7.7a). A "packet" of plasma that is emitted radially at A reaches distance #1 above the surface after a given interval of time. Meanwhile, in this same time interval, the point of emission has rotated to position B. The plasma emitted at this point travels in the next time interval to distance #1, measured from B; meanwhile, the plasma emitted at A has reached distance #2. Continuing this description for each successive time interval and connecting the corresponding positions reached by each "packet" of plasma, we see that a spiral-shaped flow line results. Emissions occur from each part of the corona. Consequently, the Sun is surrounded by a spiral-like pattern of flow lines in the equatorial plane (Fig. 7.7b), which carry the plasma out through the entire solar system.

The frozen-flux theorem of magnetohydrodynamics (Chapter 4.3) associates a magnetic field with the motion of the plasma; thus, the flow lines also describe the direction of the IMF generated by the Sun throughout the solar system. At any point, the direction of the IMF is tangential to the flow line. The polarity of the field is represented in Fig. 7.7b by solid spiral lines for a field that points north, toward the Sun, and by dashed flow lines for a field that points south, outward from the Sun. At the Earth's orbit, a southward-directed IMF can have a powerful interaction with the field lines of the magnetosphere,

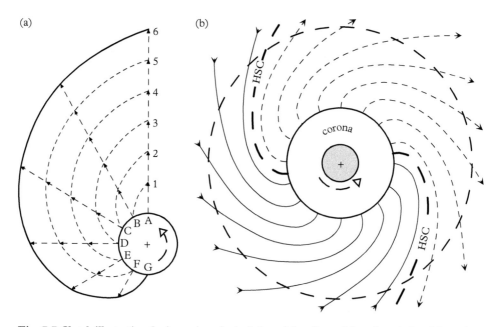

Fig. 7.7 *Sketch illustrating the formation of spiral-shaped flow lines of the solar wind and interplanetary magnetic field as a result of the rotation of the Sun during the emission of plasma.*

triggering an effect known as *magnetic reconnection*. This phenomenon is also important in the solar corona, where it triggers eruptions of plasma.

The two regions of the IMF with opposite magnetic polarity are separated by a boundary about 10,000 km thick, called the *heliospheric current sheet* (HSC), in which a very weak current of about 10^{-10} A/m^2 flows. The polarities on each side of the HSC change every 11 years at the maximum of the solar cycle. The shape of the current sheet is more complex in three dimensions, due to the variations in solar activity and associated plasma emissions. The Sun's magnetic field is dipolar only at times of minimum activity, when its axis is tilted to the axis of rotation; at other times, the field is more complex than a simple dipole. As a result of the variability, the HSC acquires an undulating shape (Fig. 7.8). It extends far beyond the planets to the limit of the heliosphere and therefore has an enormous size; it is the largest structure in the solar system. During an orbit of the Sun, each planet passes into and out of the positive and negative magnetic field regions.

Fig. 7.8 *Wave-like appearance of the heliospheric current sheet. Its massive size is emphasized by its relationship to the orbits of the planets Mercury, Venus, Earth, Mars (visible as small dots), and Jupiter.* (Image credit: *NASA/Goddard Space Flight Center. https://www.nasa.gov/content/goddard/heliospheric-current-sheet*)

7.7   Coronal Mass Ejections and Solar Flares

The difference in rotation rates between polar and equatorial latitudes in the convective zone affects the Sun's magnetic field lines. It stretches and twists them, forming loops that bulge into the corona. An electrically charged particle in the solar plasma whose direction of motion is oblique to a field line is compelled by the Lorentz force to spiral about the field line. However, if the particle's direction is parallel to a field line, it is not affected by the Lorentz effect. This means that a particle that moves along a magnetic field line encounters no resistance. As a result, loops in the magnetic field lines act as conduits for the solar plasma, carrying it outside the Sun's surface. A *solar prominence* (or filament) may form, which is a loop of plasma attached to the field lines. It may extend outward from the photosphere for tens or hundreds of thousands of kilometers into the corona (Fig. 7.9). Large amounts of energy are stored in these coronal loops. In the same way that elastic energy builds up with increasing strain when a rubber band is stretched or twisted, the energy stored in the distorted solar magnetic field lines increases until the loop becomes unstable and a snapping point is reached. This happens when magnetic field lines with opposite directions come so close to each other that they rearrange in a simpler configuration, which results in *magnetic reconnection*. It disconnects the plasma contained in a loop of field lines from the rest of the corona and releases the accumulated energy as a sudden eruption of particle radiation, that is, a *coronal mass ejection*. This may be preceded by a sudden short-lasting brightness on the Sun, that is, a *solar flare*.

Solar flares are a feature of the photosphere, chromosphere, and corona. They occur near sunspots where the solar magnetic field is locally very strong. In the corona, the local fields store energy that may be suddenly released in just a few minutes as

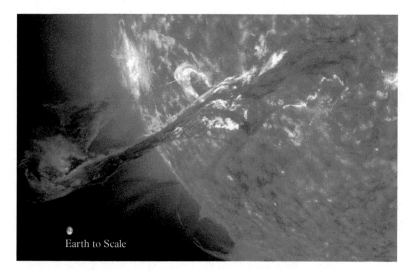

Earth to Scale

Fig. 7.9 *Eruption of a filament of plasma from the Sun's corona on August 31, 2012. The image of the Earth provides a scale for the size of the coronal mass ejection.* (Image credit: *NASA/Goddard Space Flight Center/Solar Dynamics Observatory.* https://svs.gsfc.nasa.gov/11095)

electromagnetic radiation with frequencies that cover the entire spectrum (Fig. 7.8). Only the narrow range of wavelengths in the visible part of the spectrum is evident in the short, bright flare.

In a coronal mass ejection (CME), vast amounts of magnetized plasma are expelled from the corona into the solar wind (Fig. 7.9); the average amount is in excess of a billion (10^9) tons, but occasionally much more is ejected. The ejected material usually travels to the Earth at the same speed as the solar wind, and therefore reaches it in about 2–6 days. However, some CME are ejected at speeds that exceed 3,000 km/s and reach the Earth's orbit much more quickly. The transit time to the Earth's orbit at this speed is around 14 hours.

The relationship between solar flares and coronal mass ejections is not fully understood. Both features have a strong influence on the Earth's external magnetic field. The electromagnetic radiation of a flare enhances the ionization of the atmosphere and modifies the electrical current systems in the ionosphere. The particle radiation in a CME increases the amount of plasma that enters the magnetosphere, where it can bring opposingly directed magnetic field lines so close to each other that reconnection occurs, thereby enhancing the aurorae and disturbing the magnetic field.

The CME causes large, random surges in the field measured on the Earth's surface and from orbiting satellites. The disruption is called a magnetic storm, and is characterized by rapid changes in amplitude amounting to hundreds or even thousands of nanotesla. The field surges can induce currents in exposed electrical infrastructures, such as power transmission lines, causing severe damage to terrestrial power grids, computers and electronic components. The damage that would result from a large CME striking the modern Earth would be colossal. Insurance companies have estimated that the cost of a present-day "Carrington" event would amount to trillions of dollars in the United States alone.

CMEs are not rare events. They occur with a frequency that varies from a single event every few days to a few events per day, depending on the phase of the Sun's internal activity. The frequency is thus controlled by the solar cycle. The trends in sunspot frequency can indicate when they are most likely, but the occurrence of a dangerous large event cannot be predicted exactly. In 2012 the Sun emitted a huge coronal mass ejection, similar in size to the Carrington event of 1859, that traveled in the direction of the Earth. Fortunately, the Earth had passed the region of potential impact 9 days earlier, and it and the other planets were not affected by the event. The CME was, however, recorded by the space-based solar observatory STEREO-A (Solar TErrestrial RElations Observatory).

The STEREO mission involved the simultaneous launch of twin satellites that were launched in 2006 and guided into heliocentric orbits. The orbit of the STEREO-A (for ahead) observatory is slightly inside the Earth's orbit and thus has a slightly shorter orbital period; STEREO-B (for behind) was in an orbit just outside the Earth's, with a slightly longer period, so that the twins slowly drifted apart. The observatories were designed to give a stereoscopic view of the Sun and its activity, thus enabling the observation and study of solar storms as they travel out from the Sun into space. Unfortunately, contact with STEREO-B was lost, and the observatory was abandoned in 2018. Its twin is still performing as expected and plays an important role in predicting space weather.

8

The Magnetosphere and Ionosphere

Introduction

The magnetic field close to the Earth is dominated by the geomagnetic dipole, overlain by the interplanetary magnetic field (IMF). Beyond a radial distance of several Earth radii, the IMF is dominant. When the solar wind encounters the Earth's magnetic field, the electrically charged plasma becomes trapped in the magnetic field lines. The protons and electrons interact with the field to produce electrical currents in the space around the Earth. These currents act as additional, external sources of magnetic fields. As a result, the composite magnetic field outside the Earth is much more complex than a simple dipole. The solar wind deforms the magnetic field lines into an elongated shape, compressing them on the day side and stretching them out to form a long tail on the night side. The electrical currents associated with charged particles trapped by the field lines produce magnetic fields of their own that fluctuate according to the Sun's activity, with periods that range from several seconds to decades. They include daily, seasonal, and annual fluctuations as well as the 11-year solar cycle.

In this way, the Sun influences the magnetic field of the Earth. The solar particle radiation is held at a distance by the main dipole field, which thereby shields the Earth against its negative effects, protecting life and human-made infrastructure. Large emissions of charged particles from the Sun, known as coronal mass ejections (CMEs), frequently impact on the Earth, and some are able to penetrate the magnetic shield. Their influence is greatly diminished compared to what it would be if the geomagnetic field were not present.

8.1 The Magnetosphere

The solar wind reaches the thin upper atmosphere of the Earth at a distance of about 15 Earth radii on the day side. The supersonic plasma compresses the atmosphere to form a shock wave, analogous to the bow wave in front of a boat, so that it is known as the *bow shock*. The compression heats up the solar wind and also slows it down until its speed is subsonic. The collision with Earth's atmosphere diverts the solar wind, causing

The Earth's Magnetic Field. William Lowrie, Oxford University Press. © William Lowrie (2023).
DOI: 10.1093/oso/9780192862679.003.0008

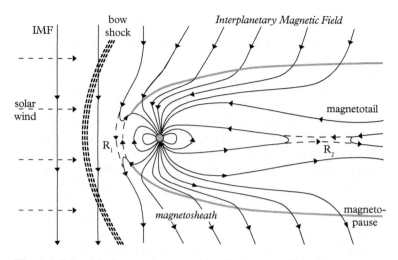

Fig. 8.1 *Schematic diagram (not to scale) to illustrate the effect of the solar wind in shaping the Earth's magnetosphere. Major features are the bow shock, magnetosheath, magnetopause, and magnetotail. Points marked R_1 and R_2 are locations of possible magnetic reconnection during coronal mass ejection events. Reconnection can occur at R_1 when the interplanetary magnetic field (IMF) is directed southward as in the diagram.*

it to flow around the Earth. At a distance of about 10 Earth-radii on the day side, the solar wind encounters the geomagnetic field (Fig. 8.1). Interactions with the solar wind distort the shape of the geomagnetic field around the planet.

The boundary between the magnetic fields of the Earth and the solar wind is called the *magnetopause*. Its closest distance from the Earth on the day side—called the stand-off distance—varies with solar activity but is on the order of 6–15 Earth radii. Between the bow shock and the magnetopause lies a turbulent region of plasma called the *magnetosheath*. The elongated region contained within the magnetopause is called the *magnetosphere*. It includes the geomagnetic field and the magnetic fields of the plasma that manages to penetrate the magnetopause. The interaction between the magnetic field carried by the plasma and the geomagnetic field results in magnetic field lines that are bunched on the day side of the Earth and are drawn out on the night side, where they form a long *magnetotail*. It extends a great distance downwind from the Earth, measuring hundreds of times the Earth's radius. Its shape has two lobes. In the northern lobe the magnetic field lines, like those of the main dipole field, are directed back toward the planet, which they enter in the northern hemisphere. The field lines point away from the planet in the southern lobe.

A coronal mass ejection (CME) that impinges on the Earth's magnetosphere compresses it additionally on the day side and draws it out to a longer tail on the night side. Where the northern and southern lobes of the tail meet, oppositely directed magnetic field lines come close to each other, creating a magnetically unstable situation that can lead to magnetic reconnection. This can also occur on the day side of the Earth. For example, if the interplanetary magnetic field carried by the solar wind is directed southward, as in Fig. 8.1, oppositely directed magnetic field lines (dashed) come close together at the point R_1 on the day side of the Earth. A similar situation is present at the point R_2 in the magnetotail. The situations at R_1 and R_2 are unstable, and additional energy from a CME can cause the field lines to break suddenly, reconnecting as closed lines (solid).

The change in energy when the field lines snap to their new positions is accommodated by injecting energetic plasma particles back into the geomagnetic field. The Earth's dipole field lines guide the charged particles down to polar regions where they collide with atoms and molecules of the upper atmosphere. The energy of the collision excites the electrons of atomic oxygen in the atmosphere and raises their energy levels to an unstable state. When the electrons return to their original lower energy state, they release the energy of the collision as light. The light emitted by an oxygen atom is most commonly green or (less often) red in color. The emissions form glowing bands around the Earth, often resembling a shimmering curtain, known as the *aurora borealis* in the northern hemisphere and *aurora australis* in the southern hemisphere. The aurorae are usually only seen in high latitudes, but the high-energy electrons in a CME can sometimes excite aurorae that are visible at sites in intermediate latitudes. At such times, the northern lights may occasionally be seen as far south as southern Europe, and the southern lights as far north as Buenos Aires.

The influx of plasma into the magnetosphere is extremely variable, reflecting the variability in the speed and density of the solar wind. Inside the magnetosphere, the plasma fluctuations produce magnetohydrodynamic waves. These consist of (1) pressure changes (compressions and rarefactions) in the plasma that propagate as *magnetosonic* waves, and (2) oscillations on the magnetic field lines that propagate as *Alfvén waves* (Chapter 7.5). The disturbances create temporal changes in the magnetic field called *geomagnetic pulsations* that range in period from a few tenths of a second to many minutes. Compared to other features of the magnetic field, these are high-frequency changes; paradoxically, they are classified as ultra-low-frequency (ULF) features in the spectrum of electromagnetic waves. At the Earth's surface, the pulsations have amplitudes between a few tenths of an nT and several hundred nT. They are classified in two groups: a group of continuous pulsations that is related to the steady solar wind and a group of irregular pulsations that occurs when magnetic storms and substorms are precipitated.

8.2   The Van Allen Radiation Belts

Protons and electrons from the solar wind that penetrate into the magnetosphere become trapped by the geomagnetic field lines as a result of the Lorentz interaction between their electrical charges and the magnetic field (Chapter 1.4). The velocity of a charged particle that encounters the Earth's magnetic field at an oblique angle can be resolved into a component that acts parallel to a magnetic field line and a transverse component that acts perpendicular to it. The Lorentz force does not change the velocity component that is parallel to the field line, but its effect on the transverse component deflects the particle so that it circles around the field line. Meanwhile, the velocity component parallel to the field carries the particle down the field line toward the Earth's surface. The combined velocity of the trapped particle forms a helix around the field line (Fig. 8.2).

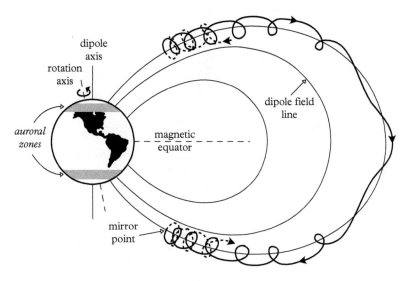

Fig. 8.2 *Sketch of the helical motion of an electrically charged particle trapped by a geomagnetic field line. As the particle moves down a field line, the pitch of the spiral becomes smaller; the particle's direction reverses at the mirror points.*

The pitch of the helix—the distance between adjacent loops—is determined by the ratio of the field-parallel component to the transverse component. As the charged particle comes closer to the Earth, the magnetic field becomes stronger, and so does the Lorentz force on the particle. This causes the transverse velocity to increase. In order for the total velocity to remain constant, the velocity component parallel to the field line must decrease. As a result, the pitch of the spiral decreases until it is zero, at which point the particle is circling the field line in a plane perpendicular to it. This happens at a place called a *mirror point*. The process then continues in the opposite direction: the helix unwinds and the particle moves back along the field line until it reaches another mirror point in the opposite hemisphere. The process repeats itself, with the particle bouncing between the mirror points in each hemisphere until it collides with other particles and loses its energy.

In this way the mirroring mechanism traps charged particles from space that enter the magnetic field and prevent them from reaching the Earth's surface. It is responsible for the existence of two belts of ions that are centered on the magnetic equator, extend about 60° of latitude on either side of it, and girdle the Earth at different altitudes (Fig. 8.3). The Van Allen belts are named after their discoverer, James Van Allen, an American space scientist. The inner belt is composed of low-energy electrons and high-energy

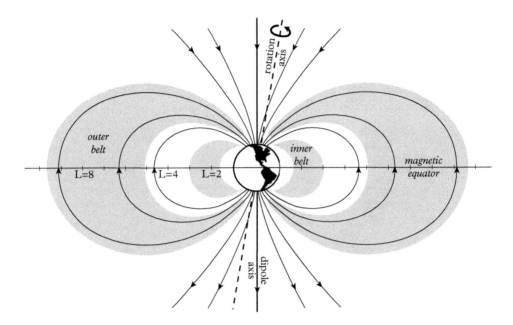

Fig. 8.3 *Schematic diagram (not to scale) showing the locations of the inner and outer Van Allen radiation belts in relation to the Earth. The boundaries of the belts are not as distinct as in the sketch.*

protons. Part of the proton population may come from the solar wind and part from collisions between the upper atmosphere and cosmic rays from outside the solar system. Together with low-energy electrons they form a region about 1,000–10,000 km above the Earth's surface. An outer belt lies at 13,000–60,000 km above the Earth's surface. In this belt the trapped particles are mainly high-energy electrons, as well as some protons, alpha particles and oxygen ions. The population of the outer belt is more strongly influenced by the solar wind and cosmic radiation from deep space, and is more variable than the inner belt. The particle density in both belts is very low, but their presence creates a hazard for spacecraft and astronauts that pass through them.

The positions of the Van Allen belts are conveniently described by the L-shells they occupy. The concept of *L-shells* is used in space physics to describe sets of magnetic field lines that occupy similar regions of space around the Earth. Consider the trajectory of a dipole field line that leaves the Earth at a given magnetic latitude in the southern hemisphere and returns to the Earth at an equivalent northern latitude. The inclination of a dipole field line at the surface of the Earth varies with latitude. Equation (3.1) gives the radial and azimuthal components of a dipole field—B_r and B_θ, respectively—at a radial distance r and azimuthal angle θ to the dipole axis (the coordinate θ is equivalent to the co-latitude). The dipole field components also determine the distance from the center of the Earth to the place where any field line intersects the magnetic equator. Let this distance be L times the Earth's radius R. The field line is identified by its L-value, which has the equation

$$r = LR \sin^2 \theta \qquad (8.1)$$

In reality, an infinite number of field lines with the same L-value form a shell around the dipole axis, all of them intersecting the magnetic equator at the same L-value and forming what is called the L-shell. Eq. (8.1) shows that the foot of a field line with $L = 6$ is at a latitude of 66°; the foot of the shell with $L = 10$ is at latitude 72°. The sets of field lines with large L-values enter and leave the Earth in polar latitudes.

The inner Van Allen radiation belt occupies the region of shells from approximately $L = 1.2$ to $L = 2.5$. The outer belt is found between $L = 3$ and $L = 10$ but is most intense around L-values of 4 to 5. The margins of the belts are diffuse and not as sharply defined as the numerical L-values might suggest. The region between the inner and outer belts is characterized by lower particle density and less intense radiation. It is known as the "safe zone" because it is relatively less hazardous to satellites, spacecraft, and astronauts.

A further consequence of the motions of the particles trapped in the geomagnetic field is the production of a *magnetospheric ring current*. This is an electric current in the magnetosphere that flows in a toroidal band around the Earth, centered on the equator, at altitudes of approximately 13,000–60,000 km ($L = \sim 2$–9), that is, in the outer Van Allen belt. The current consists mainly of protons from the solar wind, and positively charged oxygen ions from the ionosphere. It is due to a Lorentz force that arises from the gradient and curvature of the dipole field lines around which trapped charged particles gyrate (Fig. 8.2). As a particle moves along a field line, its velocity at any point has a component tangential to the line and a component normal to it. The curvature of the field line means that the tangential velocity is constantly changing direction, so that there is a slight but finite angle between the tangential velocity and the field. This gives rise to a further Lorentz force on the particle, which causes protons and positive ions to drift westward around the Earth and electrons to drift in the opposite direction. The charged particles produce a ring current that flows around the Earth in the direction of the proton drift. Viewed from above the north pole of Earth's rotation, this is a clockwise flow of positive charge. In compliance with the right-hand rule, which defines the direction of a magnetic field produced by a current loop (Chapter 1.4), the ring current causes a global field that is symmetric about the axis of the main dipole field and acts in the opposite direction to it.

Variations in the Sun's activity cause fluctuations in the intensity of both the solar wind and interplanetary magnetic field, which affect the magnetosphere and ionosphere and modulate the supply of ions to the ring current. Coronal mass ejections cause rapid enhancement of the solar wind. Magnetic reconnection between the interplanetary magnetic field and the magnetosphere feeds energy into the ring system and gives rise to magnetic storms. The changing field produced by the magnetospheric ring current induces electrical currents in the crust and mantle. Secondary magnetic fields produced by these induced currents can be measured at the surface and from satellites. Scientists analyze the secondary fields to derive information about the electrical conductivity of the deep interior of the Earth.

8.3   The Ionosphere

The Earth's atmosphere does not have a sharp upper boundary. Its density decreases gradually with increasing altitude, eventually merging into the near vacuum of outer space. Different regions of the atmosphere are defined on the basis of distinctive features they possess (Fig. 8.4). The *troposphere* is the part of the atmosphere nearest to the surface, in which all of the Earth's weather is produced. It extends from the surface to an altitude of about 10–12 km, where the flight paths of commercial jet aircraft are concentrated. The *stratosphere* is defined as the region between the troposphere and the overlying *mesosphere*, which reaches a height of about 60 km. The upper atmosphere above this height, and up to about 600 km altitude, is known as the *thermosphere*. It extends as far as the inner Van Allen belt and includes the *ionosphere*.

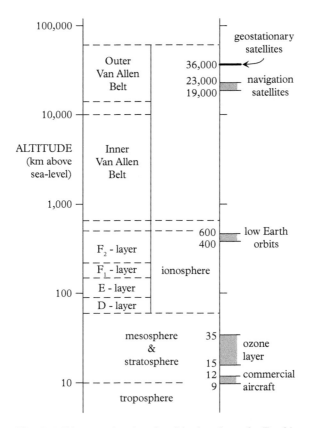

Fig. 8.4 *Diagram showing the altitudes above the Earth's surface of the upper atmosphere, the layered ionosphere, and the Van Allen belts. The positions of satellite orbits, the ozone layer, and the altitudes favored by long-distance commercial aircraft are indicated for comparison.*

The ionosphere has important effects on the geomagnetic field at the surface and in near-space. It is produced mainly by short-wavelength electromagnetic radiation from the Sun. The short ultraviolet (UV) and X-ray wavelengths of solar radiation have enough energy to be able to ionize atoms and molecules of air in the thin upper atmosphere. This happens when a photon of UV light or X-radiation collides with an electron in a neutral atom of gas and transfers enough energy to separate the electron from its atom. The removal of the negatively charged electron converts the atom to a positively charged ion. The electrons that are set free have a high energy. The low density of the outer atmosphere enables them to avoid recombination with positive ions for a short time, so that negative electrons and positive ions coexist as a plasma. This creates a shell of electrons and ions around the Earth at altitudes of 60–800 km.

Although the ionosphere is a continuous ionized region, several peaks in ionization define a quasi-layered structure. Successive shell-like layers are labeled D, E, and F in increasing order of height above the surface (Fig. 8.5). The D and E layers are most evident on the day side of the Earth where the sunlight is most intense; at night their ionization decreases so that they almost disappear. The F layer extends from 150 km to about 800 km and contains the highest concentration of electrons in the ionosphere. It is present on the day side of the Earth as a double layer, with sublayers labeled F_1 and F_2 that weaken and merge into a single F layer, which continues around the night side. The different layers of ionization act as reflectors for different wavelengths of radio waves. Shortwave radio makes use of the reflections between the ground and the ionosphere to transmit communications for distances of several thousand kilometers.

The electrically conducting plasma in the ionosphere is moved around by thermal and tidal effects. Solar heating excites the ionosphere on the day side, increasing the ionic density and forming a clearly layered structure, which partly disappears on the

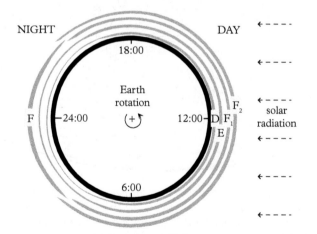

Fig. 8.5 *The layered structure of the ionosphere. The layers D and E, which are prominent on the day side of the Earth, are very weak on the night side and may even disappear.*

night side. The lunar and solar tides cause vertical displacements of the plasma, with the dominant tidal effect being the twice-daily lunar tide. The thermal expansion and tidal forces drive the plasma through the geomagnetic field lines. As a result of the Lorentz effect, the interaction creates electrical fields that produce currents in the ionosphere, mainly at altitudes of 100–130 km in the E layer.

The current system in the E layer has been investigated by systematic measurements of the magnetic field at more than 150 geomagnetic observatories around the world on days when solar activity is low (called *solar-quiet days*, or S_q-days). The magnetic field on S_q-days varies with both latitude and local time. Analysis of the variation reveals the pattern of current flow in the E layer that causes it. Viewed from above the equator on a day near an equinox, when the Sun is overhead, the current pattern consists of two large vortices on the day side of the Earth. One vortex in each hemisphere is centered at approximately 30 degrees north and south of the equator (Fig. 8.6), while weaker, oppositely polarized vortices complement them on the night side. The ionospheric currents, in their turn, produce magnetic fields that are observed at the Earth's surface, and so the ionosphere is home to an *ionospheric dynamo*.

The current vortices in the E layer maintain a steady position relative to the Sun, but, as the Earth rotates beneath them, the amplitude of the geomagnetic field at an observatory is modulated. This is known as the diurnal (or daily) variation of the field

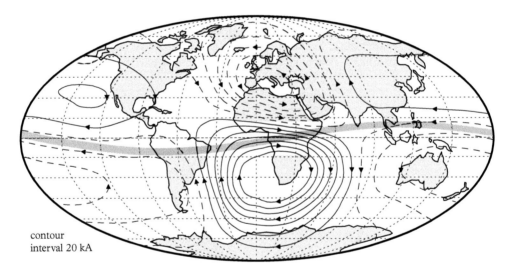

contour
interval 20 kA

Fig. 8.6 *Streamlines of large-scale electrical current circulation in the ionosphere at noon (universal time) on a quasi-equinoctial solar quiet-day (March 16, 2011). Solid contours indicate clockwise current flow; dashed contours indicate anticlockwise flow. The equatorial electrojet flows in the shaded band of latitudes within ±3° of the equator.* (Redrawn after: *M. Guzavina, A. Grayver, and A. Kuvshinov, 2018. Do ocean tidal signals influence recovery of solar quiet variations?* Earth, Planets and Space, *70:5.* https://doi.org/10.1186/s40623-017-0769-1. *Creative Commons Attribution 4.0 International License,* http://creativecommons.org/licenses/by/4.0)

(Fig. 8.7). Its amplitude is small, on the order of 10–50 nT, and depends on the latitude of the observation point. It is a tiny fraction of the intensity of the dipole field at the magnetic equator, which measures around 30,000 nT. Despite its small amplitude, the *diurnal variation* is much greater than the sensitivity of modern magnetometers, which can be better than 0.1 nT. However, the diurnal variation can exceed the amplitude of small but important magnetic anomalies related to deep geological structures and thus must be compensated during a magnetic survey.

At the magnetic dip-equator, the field is horizontal and northward. The eastward flow of ionospheric current across the northward directed magnetic field lines sets up a complex secondary system of currents in the E layer that strongly reinforces the eastward current. The additional eastward current in the E layer is called the *equatorial electrojet*. The analysis of magnetic data from the CHAMP satellite shows that the electrojet is located above the magnetic equator and is confined to a narrow band of latitudes extending a few degrees to the north and south of it. The electrojet has the effect of increasing the amplitude of the diurnal variation in the northward component at locations close to the magnetic equator. This can be seen in Fig. 8.7, where the amplitude at the equatorial site at Davao in the Philippines (station DAV) is twice as large as at nearby observatories.

Similar currents are observed in high latitudes in both the northern and southern hemisphere, but they have a different origin. As described earlier, the motion of a charged particle that enters the geomagnetic field is influenced by the angle that its path makes with the magnetic field lines. In particular, if the particle enters the geomagnetic field in a direction exactly parallel to a field line, it does not experience a Lorentz force. The particle velocity along the field line is not reduced by transfer of energy to a field-orthogonal component. The magnetic field line in this case acts as an excellent conductor that transports charged particles deep into the atmosphere, connecting the magnetosphere to the ionosphere. It gives rise to a system of *field-aligned currents* that carry electrical charges to-and-fro between the magnetosphere and the ionosphere.

Electrical charges from the magnetosphere feed the field-aligned currents at great altitudes of several Earth radii. As a result, the currents normally flow in *L*-shells with large *L*-values and affect the ionosphere in polar latitudes. The field lines allow current to flow both inward and outward between the magnetosphere and ionosphere. The inward flow is connected to the outward flow by horizontal currents across the ionosphere.

The Lorentz force has a further effect on the ionosphere. When an electrical current flows across a magnetic field, the Lorentz force deflects the positive ions in one direction and the negative ions in the opposite direction. The separation of the charges creates a secondary electrical field at right angles to both the original current and the magnetic field. This is called the *Hall effect*. In the ionosphere, it results in ring-shaped horizontal currents called the *auroral electrojets*, which flow around the magnetic poles in each hemisphere at an altitude of approximately 100–150 kilometers, that is, in the ionospheric D and E layers. During disturbed periods, when high solar activity results in an increased flow of field-aligned current between the magnetosphere and the ionosphere, the auroral electrojets increase in strength and size, and extend to lower latitudes. The magnetic fields associated with the electrojets modulate the geomagnetic field on a regular basis.

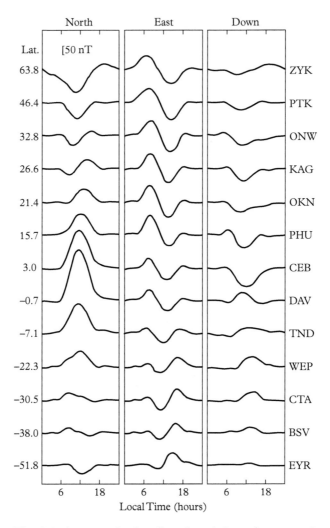

Fig. 8.7 *Average quiet-day diurnal variations of geomagnetic field components at different latitudes in eastern Asia during May–August 1996–2007. The observatories are identified by three-letter code names from the International Association of Geomagnetism and Aeronomy (IAGA). (*Data source: Y. Yamazaki and A. Maute, 2017. Sq and EEJ—A Review on the Daily Variation of the Geomagnetic Field Caused by Ionospheric Dynamo Currents.* Space Science Reviews, *206, 299–405. Creative Commons Attribution 4.0 International License, http:// creativecommons.org/licenses/by/4.0)*

8.4 Electromagnetic Induction in the Crust and Mantle

The internal magnetic field produced in the Earth's outer core changes very slowly on a human time scale. For example, since 1900 the intensity of the dipole field has decreased by about 20 nT per year at the Earth's surface, equivalent to around 5% per century. By contrast, the magnetospheric and ionospheric magnetic fields change rapidly, on time scales from seconds to years. They act as natural sources of electromagnetic induction (Chapter 1.4), causing electrical currents to flow in the crust, lithosphere, and mantle. The induced currents, in turn, produce secondary magnetic fields, which, along with the external fields, are recorded at magnetic observatories on the surface and also during dedicated magnetic satellite missions in low-Earth orbits, such as the CHAMP and Swarm projects.

The external fields originate far above the Earth's surface. The ionospheric fields result from currents at 100–130 km altitude, while the magnetospheric current system is produced far above the ionosphere, at altitudes of several times the Earth's radius. Both sources produce external magnetic fields at the surface that measure less than a few tens of nT in amplitude and thus are 1,000–10,000 times weaker than the ambient main field. However, the external fields change comparatively rapidly and are able to induce electrical currents in the conductive layers of the Earth's crust and mantle. The "skin depth" (Chapter 1.7) at which electromagnetic induction is effective in the crust and mantle—that is, the depth of penetration of the external field fluctuations—depends on the conductivity of the rocks and minerals and on the frequency of the field variations (Eq. 1.6). The lower the conductivity and/or the frequency, the deeper the penetration of the fluctuating field into the Earth.

Compared to good electrical conductors such as metals under normal ambient conditions (e.g., for copper, $\sigma \sim 6\times10^7$ S/m; for iron, $\sigma \sim 1\times10^7$ S/m), the rock-forming silicate minerals are poor conductors ($\sigma \sim 1$ S/m). There is a large variation in conductivity between different rock types, strongly influenced by their water content. By contrast with igneous rocks, sedimentary rocks are often highly porous. When moisture occupies the voids between grains, sedimentary rocks are better conductors ($\sigma \sim 10^{-3}$ S/m) than igneous rocks ($\sigma \sim 10^{-4}$ S/m). However, metallic ore minerals such as copper and nickel often occur in volcanic dikes and other structures, which results in local anomalous regions that have relatively high conductivity (e.g., in orebodies $\sigma \sim 10^4$ S/m). Electromagnetic prospecting is thus a favored method of locating orebodies.

The ionospheric magnetic field is incident as a long-period electromagnetic wave on the Earth's surface, where part of its energy is absorbed and part is reflected. The reflected wave travels back to the ionosphere, where it is again reflected. After multiple reflections, the direction of the ionospheric field is largely vertical. The analysis of both the electrical and magnetic fields created in this process forms the basis of the magnetotelluric method of electrical exploration of the Earth's crust, which has an important use in the commercial search for valuable mineral resources. It makes use of the fact that the rocks in the Earth's crust are composed largely of silicate minerals and are generally poor conductors, unless they contain veins of electrically conducting minerals or liquids such as water. In a more general research context, the electromagnetic induction caused by the electrical currents in the magnetosphere and ionosphere is an important source of information about electrical conductivity in the crust as well as in deeper regions of the crust and mantle.

It is not the only source, however. The oceans also produce electrical currents in the crust but by a different mechanism. Only about 30% of the Earth's surface is covered by land, while oceans cover the remaining 70%. The salt content of seawater makes it a comparatively good conductor ($\sigma \approx$ 3–6 S/m), depending on its salinity and temperature. The seawater is in constant motion due to ocean currents and the tides, and it is in physical contact with the crust. The oceans therefore provide an alternative method of estimating electrical conductivity that is complementary to the induction processes.

Oceanic currents are caused by differences in temperature, density, and salinity, as well as by wind systems. They are affected by the Earth's rotation and by the topography of the ocean basins. They also accompany the large-scale global displacements of seawater that accompany the tides. These result from the mutual rotation of the Earth and the Moon around their common center of gravity. Seawater accumulates on opposite sides of the Earth on the Earth–Moon axis at the expense of intermediate positions from which it is drawn away. Each day the planet rotates on its axis, passing beneath each high and each low, so that the ocean surface experiences two tidal cycles per day. This displacement is called the semidiurnal tide, denoted M2, which has a period of 12 h 25 min. The plane of the Moon's orbit is tilted to the rotation axis, which makes the semidiurnal tides unequal in amplitude. The difference has a daily, or diurnal, period.

The tidal displacements transport the electrically conducting seawater through the field lines of the ambient geomagnetic field. The Lorentz effect takes effect: the relative motion between the conductor and the magnetic field produces an electrical field, which drives electrical currents in the seawater. In contrast to the induction effects between the magnetosphere, ionosphere, and the Earth, the seawater is in direct contact with the crust. As a result, the electrical fields in the oceans produce electrical currents in the crust by a galvanic mechanism, as if the oceans were acting as a giant battery.

The diurnal and semidiurnal tides have periods similar to components of the ionospheric S_q-fields. However, analysis of long records of magnetic records makes it possible to separate the tidal and ionospheric inducing fields. The amplitude of the magnetic field resulting from the tidal motion is less than 5 nT at sea level and 2 nT at the altitude of a low-Earth satellite. Although the tidal magnetic fields are small, it is possible to use them in conjunction with larger magnetospheric-induced fields to derive a vertical profile of the electrical conductivity in the crust and mantle.

Magnetic data from several years of continuous measurements at observatories and contemporaneous long-term measurements from satellite missions have been combined to obtain a picture of the variation of electrical conductivity with depth in the lithosphere and mantle (Fig. 8.8). The tidal, ionospheric, and magnetospheric fields have periods from a few hours to a few years, which permits analysis of the electrical conductivity in depths between 200 km and 2,000 km. The magnetospheric and tidal signals were analyzed separately and jointly. On their own, the magnetospheric data were unable

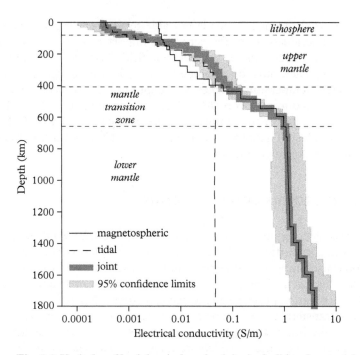

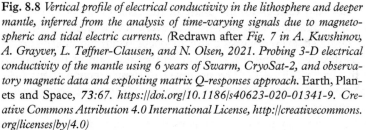

Fig. 8.8 *Vertical profile of electrical conductivity in the lithosphere and deeper mantle, inferred from the analysis of time-varying signals due to magnetospheric and tidal electric currents. (Redrawn after Fig. 7 in A. Kuvshinov, A. Grayver, L. Tøffner-Clausen, and N. Olsen, 2021. Probing 3-D electrical conductivity of the mantle using 6 years of Swarm, CryoSat-2, and observatory magnetic data and exploiting matrix Q-responses approach. Earth, Planets and Space, 73:67. https://doi.org/10.1186/s40623-020-01341-9. Creative Commons Attribution 4.0 International License, http://creativecommons. org/licenses/by/4.0)*

to distinguish the boundary between the lithosphere and asthenosphere in the upper mantle. The periods of the magnetospheric fluctuations are from a few days to a few months, which are too long to resolve the fine structure. The tidal response, produced through direct contact with the flow of seawater and dominated by the semidiurnal period, was more sensitive to the shallower structure. It resolved the increase in conductivity at the lithosphere–asthenosphere boundary and defined the conductivity profile down to depths of around 300 km. This is also evident in the joint inversion of the magnetospheric and tidal signals, which were able to resolve the conductivity profile in the mantle transition zone and to depths reaching far into the lower mantle. Below 700 km depth, the electrical conductivity ($\sigma \approx$ 1–4 S/m) is higher than in the lithosphere and upper mantle and climbs gently with increasing depth.

Induction by external magnetic fields is not able to sample deeper regions of the Earth. The electrical conductivity of the core is an important parameter in modeling the dynamo process that generates the geomagnetic field. However, it cannot be measured directly, and estimates must be obtained from models. In modeling experiments that assumed different temperature profiles with depth in the outer core and different compositions of the liquid iron–silicon alloy, electrical conductivities of $\sigma \sim 1.35 \times 10^6$ S/m have been estimated for the top of the outer core, rising to $\sigma \sim 1.55 \times 10^6$ S/m close to the inner core boundary. These values compare to $\sigma \sim 10^7$ S/m for the conductivity of iron under normal ambient conditions.

8.5   Magnetic Storms and Substorms

The amount of plasma entering the magnetosphere from the solar wind varies with the strength of the wind. In particular, the abrupt increase in plasma emitted in a coronal mass ejection causes increased penetration of the magnetopause by highly energetic protons and other particles. This produces a rearrangement of magnetic field lines in the magnetosphere, prompting magnetic reconnections. When a magnetic reconnection occurs, it transfers a large amount of energy into the magnetosphere. This produces exceptionally bright aurorae in polar latitudes and causes dramatic changes in the magnetic field at the surface, which are known as *magnetic storms*. They occur with a frequency that varies with the phase of the solar cycle, occurring once or twice a month during maximum solar activity and every few months when solar activity is minimum. The most severe magnetic storms occur when the interplanetary magnetic field (IMF) is directed southward, opposite to the direction of the dipole field on the day side of the Earth (Fig. 8.1). The northward-oriented geomagnetic field lines then join to the southward-oriented IMF lines, disconnect from the Earth, and are carried by the IMF into the magnetotail.

A magnetic storm may last hours or days. It modifies the geomagnetic field, especially the northward and horizontal components, which are reduced in amplitude. This is due to enhancement of the ring current by the increase in electrical charges in the magnetosphere and ionosphere. The magnetic field of the ring current is horizontal and southward; thus, an increase during a magnetic storm strongly diminishes the horizontal (H) and north (N) components of the main field. The reduction in H is known as the *disturbance field* and is designated Dst. In the hours following the onset of a magnetic storm, the intensity of the horizontal field drops by up to several hundreds of nT. The field remains noisy and recovers slowly over several days (Fig. 8.9). While a magnetic storm is in progress, the ionosphere is disturbed, interrupting shortwave radio communication; the irregular and unpredictable changes in intensity make it impossible to carry out a magnetic survey.

A similar but shorter-lasting disturbance of the magnetosphere is known as a *magnetic substorm*. The growth and decay of a magnetic storm last for several days and have a global effect because it injects ions into the magnetospheric ring current, some

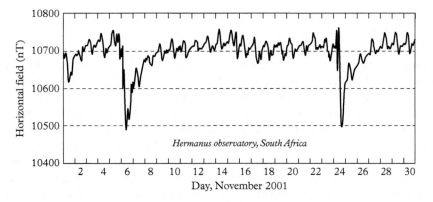

Fig. 8.9 *Variation of the horizontal component of the geomagnetic field at the Hermanus Observatory, South Africa, in November 2001, showing two magnetic storms. A diurnal variation with amplitude of ~50 nT is superposed on the main field of ~10,700 nT, which is depressed on November 6–8 and November 24–26 by magnetic storms with amplitudes of ~200 nT. (Redrawn after: Y. Yamazaki and A. Maute, 2017. Sq and EEJ—A Review on the Daily Variation of the Geomagnetic Field Caused by Ionospheric Dynamo Currents. Space Science Reviews, 206, 299–405. http:// creativecommons.org/licenses/by/4.0)*

15,000–60,000 km above the Earth's surface. By contrast, a substorm lasts only a few hours and primarily affects the polar regions, where it produces intense aurorae and large magnetic disturbances up to 1,000 nT in amplitude. The magnetic substorm is produced by enhanced currents in the ionosphere, at an altitude of only 100 km or so, much closer to the ground than the ring current. As a result, the substorms are more intense and have stronger magnetic fields than a full magnetic storm. Both storms and substorms affect the external magnetic field observed at the Earth's surface and induce currents in the solid Earth.

8.6 Space Weather

The products of enhanced solar activity—sunspots, strong solar wind, solar flares, and coronal mass ejections—have effects on the Earth and its environment that are referred to collectively as *space weather*. They cause changes in the magnetosphere, enhance the ionization of the ionosphere, and modify the magnetospheric ring current. The negative effects of space weather can damage power grids and interrupt communications on the Earth's surface. In space, there is less protection from the geomagnetic field and consequently greater exposure to space weather. The impact of energetic protons on spacecraft can disturb their performance by damaging on-board electronic components, while the increased exposure to radiation on manned spacecraft can endanger space crews and astronauts. During an extreme storm, the crew and passengers in high-flying aircraft may for the same reasons be exposed to an elevated health risk.

Major infrastructure is also at risk. The supply of electrical power to the population of a country often involves international cooperation. In Europe, for example, some countries still make electricity by burning coal or gas, which contaminate the atmosphere, or in nuclear plants, which do not contaminate immediately but which create the long-term problem of nuclear waste disposal. In the transition to an ecologically cleaner power supply, a problem arises because it is generated from wind and solar sources, that are often far from where it is to be used. Countries trade with each other, selling and buying various mixes of clean and "dirty" electrical energy. The product is transferred between countries over high-voltage power lines, which are vulnerable targets during magnetic storms. The changing magnetic field in a storm induces currents in the cables, causing power surges that can damage the transformer stations where the voltage is stepped down from the power grid to the voltage of a national grid. The potential danger was illustrated in March 1989 by a large magnetic storm that overloaded the high-voltage power transmission in Quebec, Canada, causing a loss of electrical power in the province that lasted 9 hours.

A coronal mass ejection can have similarly serious effects on global communications. The large injection of ions from a CME increases the ionization of the ionosphere. This interferes with communications that depend on the ionosphere to reflect short-wave radio transmissions. In addition to providing contacts between millions of amateur radio enthusiasts and support for emergency services, shortwave radio (now augmented by satellite support) is used for communications with aircraft over oceans and remote polar regions.

Satellite systems are particularly vulnerable to adverse space weather. Global satellite systems for navigation—such as the Global Positioning System (GPS) and its competitors, the Russian (GLONASS), Chinese (BeiDou), and European (Galileo) systems—depend on the use of atomic clocks to measure the time delay between transmission of a signal from a satellite to a receiver. Signals from a minimum number of four satellites are needed to allow the receiver to calculate its position accurately. An extra infusion of ions into the ionosphere, caused by a space weather event, changes the density of ions in the ionosphere, thereby delaying the travel time of a signal between satellite and receiver. This results in an error in location of the receiver, which can have serious consequences for military and commercial navigation.

In space, a satellite is more directly exposed to solar radiation than the terrestrial infrastructure. The extra electrical charges added to the solar wind by a coronal mass ejection can cause a spacecraft to become electrically charged. This can damage important components needed for its survival, such as solar panels, or sensitive instruments needed for its mission. Even when a spacecraft is well shielded, some ions may penetrate its defenses and affect on-board electronic circuits. The event can activate a switch or modify a part of computer memory, so that an unintentional signal is given. These events are referred to as single-event upsets (SEUs).

The occurrences of SEUs on board the Swarm satellites during the several years of the mission provide an interesting example of the shielding effect of the geomagnetic field against extraterrestrial radiation. The intensity of the field decreases with distance from the center of the Earth, and at the elevation of the satellite trio (450 km), the dipole component is about 18% weaker than on the Earth's surface. The positions of a satellite when it experienced an SEU show how the intensity of the geomagnetic field influences the incidence of particle radiation (Fig. 8.10). The number of registered impacts on the satellites during numerous orbits was high over polar regions, where charged particles are guided down the magnetic field lines. However, the most striking feature of the map is the high number of SEU impacts that occurred when the satellites were over the region of low field intensity known as the South Atlantic Anomaly (Chapter 2.9). The locally weakened field over this extensive anomaly allows the particle radiation to penetrate the Earth's atmosphere more easily. The enhanced radiation could present an increased health risk, in particular for spacecraft and high-altitude aircraft and their crews that pass through the region. The observation underlines the importance of the Earth's magnetic field as a shield against extraterrestrial radiation.

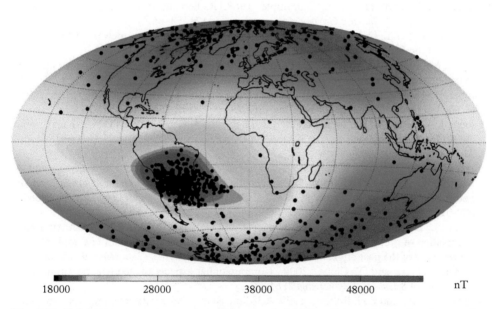

| 18000 | 28000 | 38000 | 48000 | nT |

Fig. 8.10 *Continuous tone representation of geomagnetic field intensity at 450 km altitude, according to the CHAOS-7 field model. The blue and green contours mark the South Atlantic anomaly. Black dots show the locations of the Swarm satellites when they were affected by single-event upsets due to radiation impacts. (C. C. Finlay, C. Kloss, N. Olsen, M. D. Hammer, L. Tøffner-Clausen, A. Grayver, and A. Kuvshinov, 2020. The CHAOS-7 geomagnetic field model and observed changes in the South Atlantic Anomaly,* Earth, Planets and Space, *72:156. https://doi.org/10.1186/s40623-020-01252-9. Creative Commons Attribution 4.0 International License, http://creativecommons.org/licenses/by/4.0)*

References

Bingham, D. K., and M. E. Evans. 1976. Paleomagnetism of the Great Slave Supergroup, Northwest Territories, Canada: The Stark Formation. *Canadian Journal of Earth Science*, **13**, 563–578.

Cain, J. C., Z. Wang, D. R. Schmitz, and J. Meyer. 1989. The geomagnetic spectrum for 1980 and core-crustal separation. *Geophysical Journal*, **97**, 443–447.

Cande, S. C., and D. V. Kent. 1992. A new geomagnetic polarity time scale for the Late Cretaceous and Cenozoic. *Journal of Geophysical Research*, **97**, 13917–13951.

Cox, A., R. R. Doell, and G. B. Dalrymple. 1963. Geomagnetic polarity epochs: Sierra Nevada II. *Science*, **142**, 382–385.

Cox, A., R. R. Doell, and G. B. Dalrymple. 1964. Reversals of the Earth's magnetic field. *Science*, **144**, 1537–1543.

Cox, A., R. R. Doell, and G. B. Dalrymple. 1968. Radiometric time scale for geomagnetic reversals. *Quarterly Journal Geological Society London*, **124**, 53–66.

Doell, R. R., and G. B. Dalrymple. 1966. Geomagnetic polarity epochs: A new polarity event and the age of the Brunhes-Matuyama boundary. *Science*, **152**, 1060–1061.

Domeier, M., R. Van der Voo, and T. H. Torsvik. 2012. Paleomagnetism and Pangea: The road to reconciliation. *Tectonophysics*, **514–517**, 14–43.

Dumberry, M., and C. C. Finlay. 2007. Eastward and westward drift of the Earth's magnetic field for the last three millenia. *Earth and Planetary Science Letters*, **254**, 146–157.

Dyment, J., V. Lesur, M. Hamoudi, Y. Choi, E. Thébault, M. Catalan, the WDMAM Task Force, the WDMAM Evaluators, and the WDMAM Data Providers, *World Digital Magnetic Anomaly Map version 2.0.*

Finlay, C. C., C. Kloss, N. Olsen, M. D. Hammer, L. Tøffner-Clausen, A. Grayver, and A. Kuvshinov. 2020. The CHAOS-7 geomagnetic field model and observed changes in the South Atlantic Anomaly. *Earth, Planets and Space*, **72**, 156. https//doi.org/10.1186/s40623-020-01252-9.

Genevey, A., Y. Gallet, C. G. Constable, M. Korte, and G. Hulot. 2008. ArcheoInt: An upgraded compilation of geomagnetic field intensity data for the past ten millenia and its application to the recovery of the past dipole moment. *Geochemistry, Geophysics, Geosystems*, **9**, 43–52.

Glatzmaier, G. A., and T. Clune. 2000. Computational aspects of geodynamo simulations, *Computing in Science & Engineering*, **2**, 61–67.

Glatzmaier, G. A., and P. H. Roberts. 1995. A 3-dimensional self-consistent computer-simulation of a geomagnetic field reversal. *Nature*, **377**, 203–209.

Gosling, J. T., R. M. Skoug, and D. J. McComas. 2004. Low-energy solar electron bursts and solar wind stream structure at 1 AU. *Journal of Geophysical Research*, **109**, A04104. https://doi:10.1029/2003JA010309.

Guzavina, M., A. Grayver, and A. Kuvshinov. 2018. Do ocean tidal signals influence recovery of solar quiet variations? *Earth, Planets and Space*, **70**, 5, https://doi.org/10.1186/s40623-017-0769-1.

Jackson, A., and C. Finlay. 2015. Geomagnetic secular variation and its applications to the core. In *Treatise on Geophysics*, 2nd edition, chief editor G. Schubert. Vol. 5, *Geomagnetism*, editor M. Kono, pp. 137–184. Amsterdam: Elsevier.

Jackson, A., A. R. T. Jonkers, and M. R. Walker, 2000. Four centuries of geomagnetic secular variation from historical records. *Philosophical Transactions of the Royal Society of London A*, 358, 957–990.

Kent, D. V., and M. A. Smethurst. 1998. Shallow bias of paleomagnetic inclinations in the Paleozoic and Precambrian. *Earth and Planetary Science Letters*, 160, 391–402.

Korte, M., and C. G. Constable. 2006. Centennial to millennial geomagnetic secular variation. *Geophysical Journal International*, 167, 43–52.

Kosovichev, A. G., J. Schou, et al. (34 co-authors). 1997. Structure and rotation of the solar interior: initial results from the MDI Medium-L program. *Solar Physics*, 170, 43–61, *Reviews of Modern Physics*, 74, 1073–1129

Kuvshinov, A., A. Grayver, L. Tøffner-Clausen, and N. Olsen. 2021. Probing 3-D electrical conductivity of the mantle using 6 years of Swarm, CryoSat-2 and observatory magnetic data and exploiting matrix Q-responses approach. *Earth, Planets and Space*, 7:67. https://doi.org/10.1186/s40623-020-01341-9.

Lowrie, W., and W. Alvarez. 1977. Late Cretaceous geomagnetic polarity sequence: detailed rock- and palaeomagnetic studies of the Scaglia Rossa limestone at Gubbio, Italy. *Geophysical Journal Royal Astronomical Society*, 51, 561–581.

Lowrie, W., and A. Fichtner. 2020. *Fundamentals of Geophysics*, 3rd ed. Cambridge, UK: Cambridge University Press, 419 pp.

McDougall, I., and Chamalaun, F. H. 1966. Geomagnetic polarity scale of time. *Nature*, 212, 1415–1418.

Olsen, N., G. Hulot, and T. J. Sabaka. 2015. Sources of the geomagnetic field and the modern data that enable their investigation. *Handbook of geomathematics*, 227–249, W. Freeden, M. Z. Nashed, and T. Sonar. (eds.). Berlin: Springer-Verlag, DOI 10.1007/978-3-642-54551-1.

Olsen, P. E., and D. V. Kent. 1999. Long-period Milankovitch cycles from the Late Triassic and Early Jurassic of eastern North America and their implications for the calibration of the Early Mesozoic time-scale and the long-term behaviour of the planets. *Philosophical Transactions of the Royal Society of London A*, 357, 1761–1786.

Opdyke, N. D. 1972. Paleomagnetism of deep-sea cores. *Reviews of Geophysics*, 10, 213–249.

Pitman, W. C., III, and M. Talwani. 1972. Sea floor spreading in the North Atlantic. *Geological Society of America Bulletin*, 83, 619–646.

Thébault, E., C. C. Finlay, et al. (46 co-authors). 2015. International Geomagnetic Reference Field: the 12th generation. *Earth, Planets and Space*, 67, 79. https//doi.org/10.1186/s40623-015-0228-9.

Van der Voo, R. 1990. Phanerozoic paleomagnetic poles from Europe and North America and comparisons with continental reconstructions. *Reviews of Geophysics*, 28, 167–206.

Veikkolainen, T., L. Pesonen, and K. Korhonen. 2014. An analysis of geomagnetic field reversals supports the validity of the Geocentric Axial Dipole (GAD) hypothesis in the Precambrian. *Precambrian Research*, 244, 33–41.

Vine, F. J., and D. H. Matthews. 1963. Magnetic anomalies over oceanic ridges. *Nature*, 199, 947–949.

Yamazaki, Y., and A. Maute. 2018. Sq and EEJ—A review of the daily variation of the geomagnetic field caused by ionospheric dynamo currents. *Space Science Reviews* 206, 299–405. doi 10.1007/s11214-016-0282-z

Index